Catering
for Health

Catering for Health

DANIEL R. STEVENSON
Senior Lecturer in Food and Beverage Operations,
Department of Catering Management, Oxford Polytechnic

PATRICIA M. G. SCOBIE, Ph D
Lecturer in Food Production,
Department of Management Studies for the Tourism and Hotel Industries,
University of Surrey

HUTCHINSON

London Melbourne Sydney Auckland Johannesburg

Hutchinson Education

An imprint of Century Hutchinson Ltd
62–65 Chandos Place, London WC2N 4NW

Century Hutchinson Australia Pty Ltd
PO Box 496, 16–22 Church Street, Hawthorn, Victoria 3122, Australia

Century Hutchinson New Zealand Ltd
PO Box 40–086, Glenfield, Auckland 10, New Zealand

Century Hutchinson South Africa (Pty) Ltd
PO Box 337, Bergvlei, 2012 South Africa

First published 1987
© Daniel R. Stevenson and Patrica M. G. Scobie 1987

Typeset in 10 on 12pt Palatino by
D. P. Media Limited, Hitchin, Hertfordshire

Printed and bound in Great Britain by Mackays of Chatham

British Library Cataloguing in Publication Data

Stevenson, Daniel R.
    Catering for health.
    1. Nutrition   2. Food——Composition
    3. Food   4. Caterers and catering
    I. Title   II. Scobie, Patricia M. G.
    641.1   TX551

ISBN 0 09 1649811

Cover photograph taken in the Department of Catering Management, Oxford Polytechnic, by Steve Maybury.

Contents

List of figures and tables

Preface

The increasing awareness of the general public about the links between diet and health has resulted in a large number of books which explain the issues and give recipes suitable for the home cook.

This book is written for those involved in the catering profession and the training of future chefs. It will also be of interest to the keen amateur cook and housewife. The introductory chapter covers nutrition in a practical form and includes details of current food-related problems in the UK. The chapter is written assuming that the reader has very little knowledge of nutrition.

Chapter 2 gives a guide to the nutrient profile of the recipe ingredients for those involved in food purchasing. Particular emphasis has been given to the fat and fibre content of foods. The added sugar content is more difficult to predict; quantitative information is not often available and the Food Labelling Regulations are not going to make either carbohydrate type or sugar content a compulsory part of the label.

The recipes are the result of several years of research, during which time they have been formulated, tested by the authors, and analysed for nutrient content by computer at the University of Surrey. The analysis is based on McCance and Widdowson's, *The Composition of Foods*, 4th edition (HMSO, 1978) with the exception of besan flour which was analysed independently. Most of the recipes are designed for 10 portions and are suitable for use by both caterers and those designing training sessions. However, the amateur or housewife only has to halve the stated quantities of most recipes to have an amount suitable for the average family. The recipe design and analysis is based on the metric system and the quantities used may be adapted for smaller or larger numbers by dividing or multiplying the recipes by an appropriate amount. When working in imperial measure it is recommended that any scaling up should initially be done in metric, then the totals converted to the nearest practical imperial weight; this avoids rounding errors and variations in yield.

The nutrient information at the start of each base recipe is a guide to the content of the dish per portion. *Detailed analysis of all the major nutrients is given for each recipe in Appendix 2.*

The authors have selected a wide range of basic preparations and dishes to show that *it is practical to produce healthy food in any part of a menu* and meals taken outside the home need not be the high fat, low fibre and high sugar combinations which are usually served. Furthermore, it should be stressed that *the ingredients have been chosen to maintain usual food costs at all levels within the catering industry.*

The long-term aim of these ideas is that they become incorporated into cookery as standard practices initiated at the start of a chef's training. The authors hope that *Catering for Health* will form one more link in the chain of innovation which many chefs have introduced into cookery in recent years.

D. R. S. & P. M. G. S.

Acknowledgements

The authors wish to thank the following for their assistance:

Mrs C. Briscoe
Mrs B. Davison
Dr P. Fellows
Dr P. R. Gamble
Dr G. Le Grys
Dr J. Henry
Mr M. Kipps
Mrs M. Price
Mr S. Soames
Ms J. Thorn

The authors and publisher are grateful to the following for assistance and permission to use material, tables and diagrams:

Department of Heath and Social Services
The Health Education Council
The Meat and Livestock Commission
Dr D. A. T. Southgate and Mr R. Foulkes, Food Research Institute
T. R. Suterwalla and Sons Ltd, TRS Foods
New Zealand Lamb Catering Advisory Service

The authors would also like to thank Steve Maybury, of Oxford Polytechnic, for his help in taking the cover photograph.

CHAPTER 1

Food and health

Food in history

Although there is very little evidence about the diet of prehistoric man, archaeologists have dug up tools and food residues that give some idea of eating habits in the world around 500,000 BC. Our earliest ancestors relied upon hunting and fishing for their food supplies. They ate what they could catch or find growing wild and had to expend a great deal of energy on gathering sufficient food to survive. The plant foods were unprocessed and, until the discovery of fire, all food was eaten raw.

Man's capacity to light fires was an important development as it meant that foods could now be roasted or boiled. This was the beginning of cooking which, as well as improving texture and flavour, increased the value of food, making it more digestible. If you think of the comparison between eating a raw and a cooked potato, you will have some idea of the impact that fire must have had on our early ancestors.

The next major influence on eating patterns was the advent of farming and settled community life—a process that happened at different times throughout the world. Seed gathering was replaced by cultivation and some of the wild animals were tamed. The sheep and the goat are thought to have been amongst the first animals to be domesticated—an event that also expanded the variety of foods available to include milk products. Agricultural methods slowly evolved and increased the quality of food, but the methods remained heavily reliant upon manual labour.

The general lack of transport and means of food preservation, apart from drying and salting, also contributed to eating habits. People had to use local produce for the majority of their food, much of which was only seasonally available. Detailed records of life in medieval England show that our staple food had remained essentially unchanged for thousands of years. Meals were based on vegetables and bread made from a mixture of locally ground rye and wheat flours. Fresh meat and fish were eaten in the summer but had to be salted for the winter, and those who could afford to would mask the salty flavour with spices. Honey was the main sweetener and beer the most common drink.

The change from this somewhat basic diet to that of the 1980s progressed very slowly at first. Voyagers to the New World brought back with them a trickle of unusual foods such as the potato in the late sixteenth century and tea and coffee in the mid seventeenth century. Sugar, which was initially only grown from sugar-cane and therefore imported as an expensive luxury, did not become a common household commodity until 200 years ago. This followed the discovery that sugar could be extracted from sugar-beet which will grow in northern climates. Our choice of food had begun to expand, but it is the advances in transport which came with the industrial revolution that have had the greatest impact on eating patterns. In the last hundred years the need to grow and sell food locally has completely disappeared. In the last thirty years refrigerated transport has meant that seasonal availability is becoming unimportant. We now have access to a wide range of foods from all over the world.

The way we purchase food is also changing rapidly due to advances in food technology. Raw materials are now rarely sold as grown but are preprocessed. As a minimum this involves cleaning and grading, but often will also include both preparation and cooking. In summary, eating habits have evolved very slowly over thousands of years. Changes in food availability began in earnest about 100 years ago, but they have accelerated in the last two decades, and are still continuing today.

This brief look at food in history would be incomplete without considering food requirements. We eat and enjoy twentieth century food but our bodies evolved with prehistoric man and his environment. They can adapt to changing conditions, but only very slowly. The desire to eat is as essential to survival as are the desires to drink, find shelter, or reproduce. How much food we need depends on our total activity. Table 1.1 gives a summary of these activities and how recent environmental changes can affect the need for food energy. It can be seen that the amount of energy used by modern man is less than our ancestors.

Food intake is regulated by feelings of hunger and hunger is controlled in two ways. First, on a day to day basis, by a combination of the emptiness of the stomach and the level of sugar in the blood, which drops after several hours without food. Habits, such as always eating at the same time of day, will also trigger the feeling of hunger; as can the thought of a favourite food. Second, over a period of several days intake is balanced out by needs for energy. You cannot continue to eat large quantities of food, however appealing, without losing your appetite and therefore reducing your intake. Any excess food energy is either burnt off as heat or stored as a fat layer under the skin, a useful attribute for early man whose food supplies were erratic. However, this hunger mechanism was designed to cope with unrefined food, high activity levels, and extremes of temperature. Problems with food intake in the past were usually associated with a lack of food, for example starvation, or the lack

Table 1.1 *Activities which require food energy*

Activity	Description	Change in energy used	
Basal metabolic rate	Energy for breathing, heartbeat and body function	Due to height increase	MORE
Heat production	Energy needed to maintain body temperature	Heated buildings, modern clothing	LESS
Movement	Energy needed to move about, varies with speed and body weight	Mechanized tools and transport	LESS
Growth and repair	These need both energy and food materials, vary with age and state of health	Overall improvement in health	
		Increase for growth	MORE
		Decrease for repair	LESS

of a particular ingredient such as vitamin C which leads to the deficiency disease scurvy.

Problems associated with food in present day Britain are caused in part by the changes in requirement outlined in Table 1.1, but more particularly by the dramatic changes in food variety which have occurred. Scarcity is now almost unknown, and many foods are highly refined, often to the point where it is difficult to identify their original ingredients. They are advertised and packaged in an appealing manner, are quick to digest, and can contain high levels of sugar and salt, all of which tends to blunt our appetite mechanism and leads to many of us becoming overweight.

However, there is some indication that food intakes have declined a little. There has also been a subtle change in the proportions of food that we eat. The amount of energy coming from fat has increased while the amount coming from vegetables and cereals, especially unrefined cereals, has declined. This in turn has led to a decrease in the fibre content of our food. These changes are summarized in Figure 1.1 (see overleaf).

The mix of foods that we eat no longer matches our requirements and as a result there has been an increase in ill health which is directly associated with our eating habits.

The most frequent cause of death in the UK is **coronary heart disease (CHD)**; this was uncommon before 1925, but now kills three out of ten

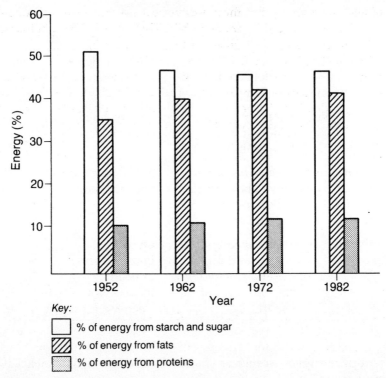

Figure 1.1   *The changes in food consumption between 1952 and 1982. From the Ministry of Agriculture, Fisheries and Food.*

men and two out of ten women. These levels are amongst the highest in the world and show little sign of declining. One of the three main risk factors for CHD is a diet that is too high in fat, especially animal fats. Hypertension (high blood pressure) is also a risk factor and this is associated with a high salt intake. The third main factor is smoking.

Cancer of the large bowel is the second largest cause of death from cancer. Low intakes of cereals and pulses and high intakes of fat, sugar and eggs are thought to be linked with this form of cancer. A fibre-rich diet also helps to prevent constipation and diseases of the colon.

High sugar intakes are one of the main causes of **tooth decay** and over 90% of children will have had some tooth problems before they leave school.

Sugar, which is so easy to eat, **fats** which are relatively high in energy, and **alcohol** contribute to the most common problem of all—**overweight**. This affects 39% of adult men and 32% of adult women. Amongst this number, there are 2.8 million (7%) of people, the obese, whose weight is at least 20% higher than an acceptable level.

It is hardly surprising that there has been a rise in concern amongst health professionals and the general public. This concern has culminated in a series of reports on the subject of food and health.

In 1978 the DHSS published *'Eating for Health'*, which gives general guidelines, and this was followed in 1983 by a report from the National Advisory Committee on Nutrition Education (NACNE). The NACNE report gives a detailed list of dietary goals for the 1980s and 1990s. It includes recommendations for the amounts of food needed and the scientific basis for the changes suggested. The specific problem of coronary heart disease was the subject of the Committee on Medical Aspects of Food Policy (COMA) report in 1984 and a subsequent booklet *'Eating for a healthier heart'*. Currently government recommendations and legislation are being drawn up to improve the way food is labelled. The common message coming from these reports is that we should:

eat less fat—cut down by 25%
eat less sugar—cut down by 50%
eat more fibre—increase by 50%
eat less salt

Catering for Health is designed to assist the Catering Profession to follow these guidelines. The remainder of this chapter will look more closely at food components. Chapter two describes food commodities and the remainder of the book gives recipes and the practical information needed to *cater for health*.

Food components and health

The food we eat is described in terms of the nutrients that it contains. There are two types of nutrient: first, those which provide energy when absorbed and broken down by the body, these are the carbohydrates, proteins, fats and alcohol; second, those nutrients that do not provide energy but are essential to body function, the vitamins and minerals. It is extremely important to obtain the right balance of nutrients, as any one type should not predominate. The majority of foods contain a mixture of nutrients and an understanding of their nature and function is essential to appreciate the links between food and health.

The carbohydrates

Carbohydrates come from plants and can be divided into two groups: those that can be broken down by the digestive system to produce energy—the **available carbohydrates**; and those that we cannot digest but which are essential as they provide bulk—the non-available carbohydrates or **dietary fibre**.

Available carbohydrates
These are subdivided into two categories.

Sugars
The sugars are carbohydrates which dissolve readily in water.

Sucrose or common sugar is the most familiar. It is widely used because of its sweet taste and the way it readily combines with other ingredients in cookery. White sugar is one of the most highly refined foods and it contains only one nutrient, sucrose. Brown sugar is little different having only traces of colouring matter and minerals in addition to the sucrose.

Lactose is the sugar in milk and does not taste sweet.

Maltose, formed from starch, is largely used by the brewing industry.

Fructose, found in ripe fruit and honey, is even more sweet tasting than sugar.

Glucose is a simple sugar which forms the basic unit of all other carbohydrates and is found in many plants, honey and in the blood. Glucose is again less sweet tasting than sugar but has water-holding properties which are widely used in the confectionery industry. Glucose syrups, for instance, keep the centre of chocolates soft. Almost all other carbohydrates are broken down into glucose during digestion.

Starch
Starch is the food store of plants and unlike the sugars it will not dissolve in water. Starches are composed of chains of glucose units and are difficult to digest until cooked. Cereals such as wheat flour and rice, the pulse vegetables and potatoes are amongst the most important sources of starch, but it is also found in a large variety of other foods from spices to unripe fruit.

Dietary fibre
This is a general term which is used for the combination of complex carbohydrates that passes through the digestive system largely unchanged. It is a mixture of **celluloses**, which are the fibrous substances that make up plant cell walls, and **pectins**, found in many fruits, some of which can be extracted to make gels. **Lignin**, a woody substance, is also included in this group although not a carbohydrate, as it is eaten with the other types of dietary fibre and is completely indigestible. Although dietary fibre passes into the lower end of the gut (the colon) unchanged, it is broken down there to some extent by fermentation; gas and a small amount of energy are produced as a result of this breakdown.

The role of carbohydrates

The carbohydrates in food have two main functions. The first is to provide a relatively cheap source of energy, and the second is to provide sufficient fibre to ensure that the digestive system functions efficiently.

Energy is measured in **kilojoules** (kJ) which are standard international units or the more familiar **calories** (kcal). The various types of sugar and starch all provide similar amounts of energy.

1 g of sugar provides 16 kJ or 4 calories of food energy
1 g of starch provides 16 kJ or 4 calories of food energy

Despite this similarity, there are very different effects on the body when sugars and starches are eaten. There are several problems associated with sugars which do not occur when starches are used as a source of energy.

Added sugar in the form of white sugar or glucose syrup makes by far the largest contribution to our total sugar intake. The lactose in milk and the fructose in fruits and honey are the only other sugars eaten in any quantity. In 1983 the average person in the UK put 16 kg of added sugar into their drinks and onto their food. But this was only 38% of the added sugar used. The remaining 26 kg was eaten in sweets, ice-cream, cakes, biscuits, tinned fruit and vegetables, chocolate and soft drinks, which are just a few of the products that contain hidden sugar. The problem with such a large intake of sugar is that it is a highly refined food containing only one nutrient, sucrose. Sugar contributes 14% of total energy needs but *nothing else*, while all other foods provide both energy and essential nutrients, such as proteins, vitamins and minerals.

Tooth decay or dental caries occurs most frequently in countries with high sugar intakes. The UK is one of the highest consumers of sugar and dental caries is present in 95% of the adult population who still have teeth left (unfortunately approximately 30% of adults are toothless). Bacteria in the *plaque* on the teeth use sugar and form acids which then erode the hard surface of the teeth and cause decay. This is most likely to happen if sugar is eaten between meals especially in a sticky form, such as toffee, which will cling to the teeth. The action of chewing fibrous foods or brushing helps to remove the plaque, while fluoride in the water supply makes teeth more resistant. Children are the most vulnerable to dental caries and are also the biggest sweet eaters.

Being *overweight* is a condition which affects all ages of the population but becomes increasingly common as one grows older; 15% of teenagers (age 16–19) are overweight and by late middle-age (age 60–65) the figure has risen to 52%. The result of being overweight is an increased risk of heart disease, arthritis and diabetes, as well as a shorter life expectancy. The heavier you are the greater the risks. The cause of overweight is a confused mixture of eating more than you require and taking too little exercise. Unfortunately some people seem more susceptible than others. Sugar adds to the problem because it is high in energy, low in bulk and therefore easy to eat in large quantities.

The NACNE report proposes that sugar intakes should be reduced to 20 kg per head per year.

This figure is about half the amount currently eaten. It is particularly important to reduce the amount taken as snacks and in soft drinks, which should be less than 1 oz or 28 g per day.

Starch, on the other hand, is a source of energy which comes in combination with the fibre of cereals, fruit and vegetables. These foods also provide protein, vitamins and minerals. Their relative bulk makes them more satisfying to eat than high-sugar foods and less likely to cause an excessive calorie intake. However, this is only true of cereals if they are eaten in the same form as they grew, because modern cereal processing refines sources of starch to reduce the fibre content. When wheat is ground into wholemeal flour, all of the grain is used and the flour contains 9.6% fibre. Refined white flour only contains 70% of the original grain as much of the outer husk is removed, reducing the fibre content of white flour to 3.4%. Some breakfast cereals are also highly refined and contain added sugar.

The combination of cereal processing and the general trend to eat less bread and potatoes means that dietary fibre intakes are too low. Fibre is important because it has good water-holding properties and therefore adds bulk to the mixture of foods passing through the gut and ensures a quick transit time.

Low fibre intakes are associated with several diseases of the colon:

Constipation is the most common problem and 22% of the population take laxatives at a cost that runs into millions of pounds.

Diverticulitis, the formation of infected pockets in the colon, is another problem which often follows constipation. Diets rich in fibre, particularly from cereals, protect against these two conditions.

Cancer of the colon, which is the second largest cause of death from cancer, is associated with low fibre and high fat intakes and it is noticeable that vegetarians have a lower incidence of colonic cancer.

The NACNE report recommends that fibre intakes should be increased to 30 g per day.

This is a 50% increase on average UK intake. To do this we will have to eat more unrefined wholegrain cereals, as well as pulses and fruit and vegetables. These foods have several advantages apart from their fibre content.

They add variety and colour to menus, take longer to chew, and their bulk gives a greater satisfaction to meals, which helps to reduce the intake of sugar and fat. The effect of refinement on the carbohydrates in foods is shown in Table 1.2.

Table 1.2 *The carbohydrate content of a selection of refined and unrefined carbohydrate foods*

	Fibre	Starch (g/100 g)	Sugars	Energy (kcal)
Refined foods				
Sugar	–	–	105	394
Chocolate	–	3	57	337
Cornflour	–	92	–	354
Spaghetti	–	81	3	378
Patna rice	2.4	87	–	361
White bread	2.7	48	2	233
White flour	3.7	78	2	337
Brown bread	5.1	43	2	223
Muesli	7.4	40	26	368
Cornflakes	11.0	77	7	368
Unrefined foods				
Brown rice	4.2	80	–	357
Wholemeal bread	8.4	38	2	216
Wholemeal flour	9.6	87	2	357
Shredded wheat	12.3	67	–	324
Puffed wheat	15.4	67	2	325

The fats

Fat is a nutrient which provides a very concentrated source of energy. The term 'fat' however is confusing as it covers both the solid fats chiefly found in animal foods and liquid fats or oils found in plants. All fats and oils are a mixture of similar compounds called the triglycerides. Each triglyceride contains one unit of glycerine and three **fatty acids**, but they differ from each other in their mixture of fatty acids. It is the nature of these fatty acids which is important both in catering and in health. There are over 40 different known fatty acids, which are divided into three groups.

Saturated fatty acids
Saturated fatty acids have a tight chemical structure which makes them stable and solid at room temperature. They make up a large proportion of

the fats in meat, lard and milk products, but are also found in plant foods, particularly coconut and palm oil.

Polyunsaturated fatty acids

Polyunsaturated fatty acids have a more open chemical structure than saturated fatty acids. They are liquid at room temperature and are more easily broken down in the presence of air when they become rancid or they smoke during frying. Their structure also makes it possible to add hydrogen to a liquid oil and manufacture a solid fat such as margarine. **Trans fatty acids** are formed during this process and have similar properties to saturated fatty acids. Two of the polyunsaturates, linoleic and linolenic acids, are said to be essential as they are required for cell structure and cannot be made in the body from other sources. Polyunsaturated fatty acids predominate in the fats of sea fish and the following vegetable oils: groundnut, corn, sunflower, rape and safflower.

Monounsaturated fatty acids

Monounsaturated fatty acids are an intermediate variety of fatty acid which are liquid at room temperature and found in a wide range of foods. Olive oil is a particularly rich source.

Any naturally occurring fat or oil is a mixture of triglycerides and will contain up to 16 different common fatty acids plus traces of many more. It is difficult to give precise figures for the amount of a fatty acid in a food, especially from animals, as the type of feed given will alter the fatty acid composition of the flesh. Babies for instance show a distinct difference in fat stores when changed from breast to bottle feeding. However, foods have a regular pattern of fatty acids, as shown in Table 1.3. The mixture of solid and liquid fatty acids in fats gives them the creaming property so often used in cookery. A solid fat will soften over a range of temperatures and allow air, sugar or flour to be blended with it.

Cholesterol

This is a substance which is similar to fats and closely linked to them because of its function. Cholesterol is required everywhere in the body as it forms part of cell structure, and is carried around in the blood stream with other fats. It is also a constituent of some hormones and of bile salts which aid fat digestion.

As with the carbohydrate containing foods, fat comes in both unrefined and highly processed forms. The unrefined sources are all meats, oily fish, egg yolk, milk and nuts; these foods also contain a variety of other nutrients, particularly protein. The refined forms are lard, margarines, cooking and salad oils, cocoa butter and the many products made with

Table 1.3 *The fatty acid composition of some common foods (% of total fatty acids)*

Food	Saturated	Monounsaturated	Polyunsaturated
Butter, milk, cream	61	30	3
Lamb	50	40	5
Beef	48	48	3
Pork and bacon	42	50	8
Palm oil	45	45	9
Margarine (hard)	40	33	13
Chicken	34	45	20
Olive oil	14	73	12
Groundnut oil	15	53	31
Fatty fish	23	27	50
Margarine (polyunsaturated)	23	22	53
Corn oil	14	31	55
Soya bean oil	14	24	60
Sunflower oil	12	33	58

these fats such as chocolate, cakes, biscuits and mayonnaise. Butter and cheese are also rich sources of fat and, although not highly refined, are a concentrated form of milk fat. However, none of these foods are pure fat.

The role of fats

Fats are essential nutrients whose chief function is the provision and storage of energy. They are a very concentrated source of energy containing more than twice the amount in carbohydrates.

1 g of fat provides 37 kJ or 9 calories of food energy

Anyone doing a great deal of physical exercise or living in a very cold climate would find it difficult to survive without a reasonably high fat intake. Energy stores are laid down under the skin in the form of fat in both humans and animals. The other important function of fats is that they are a source of essential fatty acids and fat soluble vitamins.

However, the population of Britain is not particularly energetic, nor is the climate excessively cold, but the amount of energy that we obtain from fat has been slowly increasing, as was shown in Figure 1.1. In 1950, when accurate figures were first recorded in the National Food Survey, fat contributed 36.8% of the calories. By 1984 this figure had risen to 42.3%, an increase of 16%, and nearly half of this is the saturated type. As was shown with high sugar intakes, a high fat intake can lead to major health problems.

Cardiovascular diseases

Cardiovascular diseases are heart attacks, the medical name for which is coronary heart disease (CHD), and strokes, or cerebrovascular disease, and together they account for more than half of all deaths in the UK. The incidence is especially high in Scotland and Northern Ireland and is not just confined to the older members of the population. It is estimated that there is an annual loss of a quarter of a million years of working life because of the death of men under the age of 65 in England and Wales. However, not all CHD and strokes are fatal but they do cause a varying amount of disability and unhappiness, which is also a drain on time and resources.

These two diseases develop slowly from childhood onwards with a gradual accumulation of fats and fibrous tissue in the walls of the blood vessels, which reduces their size. A heart attack or stroke occurs when one of the restricted blood vessels is suddenly blocked by a blood clot. Although there is no single cause, all the countries which have a high fat intake also have high death rates from CHD, therefore, the British government commissioned a panel of experts to advise on the significance of the links between diet and cardiovascular diseases. The result was the COMA report which made the following recommendations on fats.

The consumption of saturated fatty acids and of fat in the United Kingdom should be decreased

The specific recommendations are that not more than 35% of total food energy should come from fat and not more than 15% of food energy should come from saturated fatty acids.

In daily terms, this means a drop in average fat intake from 104 g to 82 g, a reduction of approximately 25%. Although the lower intakes of total fat will lead to some reduction of saturated fatty acids, the COMA report also recommended an increase in the proportion of polyunsaturated fatty acids in the diet. The NACNE report had already made similar suggestions with a slightly higher reduction in total fat intake to 30% of food energy.

The reason for these recommendations is the evidence that high blood cholesterol levels are strongly associated with CHD. While small amounts of cholesterol are eaten in eggs and shellfish, the majority is produced by the liver from other foods. High intakes of fat, especially if the fats contain a large proportion of saturated fatty acids, result in raised blood cholesterol levels. However, a diet containing a greater proportion of polyunsaturates tends to have the reverse effect and lower blood cholesterol. Monounsaturates appear to have no effect upon blood cholesterol levels.

The COMA report also states that:

1 Fibre-rich carbohydrates should replace the fats.
2 Sugar intakes should rise no further and an excessive intake of alcohol should be avoided as both of these can lead to obesity.
3 Obesity increases the risk of CHD and should be avoided by a combination of lower food intake and increased exercise.
4 Salt intakes are unnecessarily high (see page 23).
5 **People should not smoke cigarettes** as smoking is another major risk factor.

Unfortunately, the likelihood of a person developing cardiovascular disease multiplies when they have more than one risk factor. An overweight smoker with a high saturated fat intake is about eight times more likely to have a heart attack than a person with none of these problems.

Obesity
Anyone whose weight is 10% above the standard for their height is said to be overweight, and at 20% above they are obese. In either case, a diet high in fat will contribute to the excess weight. Fat is described as an energy dense food because it contains so many calories in a small volume of food, for example:

   1 teaspoon of oil weighs 4 g and contains 36 calories
   1 teaspoon of sugar weighs 5 g and contains 20 calories
   1 teaspoon of flour weighs 4 g and contains 14 calories

Also fat is present in easily eaten foods where its presence is not always obvious such as cakes and biscuits; milk, cream and ice-cream; and meat products like pies and sausages.

The contribution that different foods make to fat intake in the UK is shown in Table 1.4. Many of them are important sources of other nutrients and therefore care needs to be taken when choosing how to reduce fats. A combination of small changes is the safest approach and the following is a list of guidelines for caterers to use when planning meals. Full details of the different foods are given in Chapter 2.

1 Menu planning
Reduce the number of deep fried items.
Always offer an alternative to red meat.
Feature fish, chicken and vegetarian dishes on the menu.
Substitute cream in desserts with fruit and yoghurt.

2 Purchasing
Buy only polyunsaturated oils or margarine and offset the cost by using less.

Choose the lower fat variety of a food; milk, for instance, is available in semi-skimmed and skimmed forms. Meat grading schemes now offer leaner carcases and surface fat can be trimmed off joints.

Offer milk as well as cream in coffee and use half cream instead of full cream.

Check the fat content of ready prepared foods using the fat labelling regulations (see page 38).

Use in-house meat products or those with a known low fat content.

3 Cooking
Use recipes with lowest possible fat content.

Never garnish with extra fat, for example butter on vegetables.

Shallow fry in heavy duty pans with a minimum of oil.

Roast items by brushing on oil and not rolling in hard fat.

Deep fry at the correct temperature and cut chips as large as possible.

4 Personnel management
Explain to staff the reasons behind any policy changes. The Health Education Council leaflets listed on page 34 will help and remember that overweight is a particular problem in the catering industry.

Table 1.4 *The contribution of different foods to fat intake in the UK*

Food	% Total fat	% Saturated fat
Milk and cream	14.5	20.3
Cheese	5.0	7.1
Meat (fresh)	14.0	13.0
Poultry	1.4	1.0
Sausages	3.1	2.8
Meat products	6.7	6.4
Fish	1.3	0.7
Eggs	2.9	1.9
Butter	9.8	14.7
Margarine	13.9	9.8
Oils and other fats	12.1	9.8
Bread and flour	2.7	1.3
Cakes and biscuits	6.3	7.1
Cereals	1.7	1.2
Vegetables	2.2	1.2
Fruit	1.0	0.6

Source: *The National Food Survey for 1984* (MAFF, 1986).

Proteins

Proteins differ from carbohydrates and fats because they contain nitrogen and form part of the basic structure of every cell in the body. The muscles, bones and nerves are all made with a framework of protein. The enzymes and hormones which are responsible for body regulation are also proteins. Protein is therefore essential for general health.

Proteins are complex structures made up of long chains of several hundred amino acids and they are found extensively in both animal and plant foods. Each species contains protein with its own characteristic mix of amino acids. However, the body has the capacity to break down the protein chains during digestion, absorbing the resultant amino acids and rebuilding them into the type of protein needed for a particular tissue. There are at least 20 common amino acids and eight of these are essential as they cannot be made in sufficient amounts by the body but must be provided by other foods. A mixture of animal or plant proteins will provide all the essential amino acids, but a narrow choice of food can lead to one or more amino acid being deficient, as can happen with a strict vegetarian diet or during a famine.

Milk and meat have the reputation for being superior sources of protein as they contain all eight essential amino acids in the correct proportions. However, a combination of cereals and pulses with a little meat or milk will provide a similar amino acid mix at a lower cost. The nutritional advantage of animal foods has more to do with the iron and vitamin content than their protein type (see page 21). The protein content of certain foods is given in Table 1.5 (see overleaf).

The role of proteins

Proteins are the basis of body structure and are, therefore, essential for growth and it is particularly important that children and pregnant women have adequate amounts. A low protein intake in childhood results in stunted growth. This is shown clearly by the average increase in height in the UK following improvements in the general standard of living during the first quarter of this century. However, the body is continually wearing down and replacing tissue and the skin surface, making protein essential for people of all ages. Any one recovering from injury also needs a high protein intake to provide material to repair the damage.

Any excess protein eaten, but not required for growth or repair, is used as energy and the nitrogen released during this process is removed from the body in the urine.

1 g of protein provides 17 kJ or 4 calories of food energy

It is recommended by the Department of Health and Social Security (DHSS) that protein should provide 10% of a person's energy needs (see

Table 1.5   *The protein content of a selection of foods (g per 100g)*

Cheeses		*Eggs*	12.3
Cheddar type	26.0	*Fish*	
Cottage	13.4	Cod fillet	17.4
Feta	16.5	Salmon	18.4
Fromage blanc	8.4	Prawns	22.6
Meats (average)		*Flours*	
Chicken	19.7	Besan	21.0
Beef	16.6	White	9.4
Lamb	16.2	Wholemeal	13.2
Pork	16.9		
Turkey	22.0	*Nuts*	
Liver	20.1	Cashew nuts	17.8
Milks		Peanuts	24.3
Whole	3.2	*Pulses*	
Skimmed	3.3	Baked beans	4.8
Yoghurt	5.0	Kidney beans	22.1
Cereals		Chick peas	20.2
Cornflakes	8.6	Lentils	23.8
Pasta	2.0		

page 232). For example, a moderately active 35 year-old man requires 69 g of protein a day and 2750 calories. The 69 g of protein provides 276 calories (4 × 69) and this is 10% of the 2750 daily calorie requirement. However, these recommendations are known to be high as many poorer nations can remain healthy on much lower protein intakes provided their food contains adequate energy.

In 1984 proteins provided 13.1% of the total energy consumed in the average diet of a person in the UK and animal foods contributed 63% of this protein. This level is in excess of requirements but the amount of extra energy involved is small and unlikely to cause overweight. Both the NACNE and COMA reports state that protein intakes should remain unchanged but that the proportion of animal protein should decrease because of its association with saturated fat.

Alcohol

Alcohol is not usually classed as a nutrient or a food because of its habit forming nature and the hazards associated with an excessive intake. It does however provide a substantial amount of energy and cannot therefore be ignored.

1 g of alcohol provides 29 kJ or 7 calories of food energy

This is considerably more than the energy value of an equivalent weight of protein or carbohydrate. Alcohol is absorbed into the body from both the stomach and the small intestine at a variable rate. Absorption will be fast if the alcohol is drunk alone on an empty stomach, but significantly slower if taken with other foods. Once in the blood the alcohol is transported to the liver and there broken down to provide energy. The speed of this breakdown varies with different people and is related to their body weight; an average breakdown rate would be 100 mg of alcohol per kg body weight per hour. For example if a 65 kg man drinks two pints of ordinary bitter, it will take approximately 5½ hours to clear from his blood. The speed of breakdown *does not* increase with the quantity imbibed; so the length of time that alcohol remains in the blood is dependent on the amount taken in. It could take several days to breakdown the results of a heavy drinking session and if the person drinks again during that time they suffer a topping up effect on blood alcohol levels.

The effects of alcohol

Social
Alcohol acts by depressing the nervous system and thereby making people feel more relaxed and sociable. This effect has been known for centuries and is the reason why beer, wines and spirits have formed an integral part of hospitality. Taken in moderate quantities, alcohol can reduce stress and is unlikely to cause any health problems.

Skill levels
The depressant effect of alcohol reduces skill in fine movement and makes it harder to judge distances, hence the campaign not to drink and drive. One-third of the drivers killed in road accidents have blood alcohol levels that are over the legal limit.

Physical
Hangovers, or the unpleasant side-effects of drinking, have two main causes. The feeling of thirst is due to the alcohol itself, but headaches are caused by *congeners*, substances associated with alcohol and found in some drinks. The most likely sources of congeners in descending order of importance are brandy, red wine, rum, whisky, white wine and gin.

Gastric problems are caused by an excessive intake of alcohol irritating the stomach or the colon.

Health problems
Alcohol is often an underlying reason for overweight as it provides energy in addition to that coming from food.

Regular heavy drinking causes liver damage or *cirrhosis*, a condition which is much more common in countries with high alcohol consumption and which is increasing in the UK. There are also links between alcohol intake and a variety of other diseases including *hypertension*, brain damage, and malformation in babies.

Alcoholism
Alcoholism is a state of irrational dependence on alcohol which can occur after years of heavy drinking. Alcoholics find it difficult to maintain normal relationships with other people and to cope with their work. It is difficult to know why some drinkers become alcoholic, but certain groups of people are more susceptible to the disease: business executives who have the financial capacity and social pressure to drink; those with easy access to alcohol such as publicans; and the lonely. An alcoholic needs expert help to recover and will normally then have to refrain from ever drinking again. It is, therefore, very important to recognize the potential problem before a state of dependence is reached.

There has been a steady increase in alcohol intake in the UK over the last 20 years, largely due to a drop in the price relative to other foods and to the easy and inconspicuous way it can be purchased in a supermarket. Current estimates are that between 4% and 9% of total energy comes from alcohol and given the increase in alcohol related problems, this is thought to be too high. The difficulty lies in suggesting a moderate level which will allow for social enjoyment without endangering health. This is a particularly pertinent problem for those in the Hotel and Catering Profession who are both at risk personally and reliant on the profit from beverage sales as part of their income.

The amount of alcohol that it is reasonable to drink has been defined in two ways. Firstly, the Royal College of Physicians, in its report on alcohol and alcoholism, gives an amount that should not be exceeded.

Regular alcohol intakes of the equivalent of four pints of beer a day for a man should not be exceeded.

The amount for women is about half this as they tolerate alcohol less well because of their smaller frame size. Secondly, the Health Education Council has made recommendations based on keeping within safe limits.

To keep within a safe limit men should not drink more than two or three pints of beer (or their equivalent) two or three times a week. The recommendations for women are also half those for men.

On a national level the NACNE report suggests that average intakes should drop by 10%, which would mean that alcohol provides 5% of total energy.

The other important factor to note is that many drinks contain substantial amounts of sugars as well as alcohol. Spirits are the exception but they are often taken with mixers which contain sugar. Public concern about alcoholism and the campaign against drinking and driving has resulted in a rapid growth of the mineral water and soft drinks industry.

The average alcohol content and energy value of some common drinks is given in Table 1.6.

Table 1.6   *The alcohol, sugar and energy values of some common beverages*

Beverage	Alcohol (g/100g)	Sugars (g/100g)	Energy (kJ/100g)	(kcal/100g)
Mineral water	–	–	–	–
Tomato juice	–	2.3	66	16
Ordinary bitter	3.1	2.3	132	32
Lager	3.2	1.5	120	29
Mild ale	2.6	1.6	104	25
Red wine	9.5	0.3	284	68
Dry white wine	9.1	0.6	276	66
Sweet white wine	10.2	5.9	394	94
Dry sherry	15.7	1.4	481	116
Sweet sherry	15.6	6.9	568	136
Spirits (70% proof)	31.7	–	919	222

A guide to the relative content of alcohol in a standard measure of different types of drink is given in Table 1.7. This is based on a unit system, each unit containing 8–10 g of alcohol, however, it should be remembered that the strength of beers and wines is variable.

Vitamins

The vitamins are unrelated substances that are grouped together for the following reasons.

1 Vitamins are essential for body regulation.
2 Vitamins cannot be manufactured by the body and therefore must be

Table 1.7   *The relative alcohol content of some standard drinks*

1 pint of ordinary beer	= 2 units
1 glass of wine (4 oz)	= 1 unit
1 glass of sherry (2 oz)	= 1 unit
1 glass of port (2 oz)	= 1 unit
Single measure of spirits	
in England and Wales	= 1 unit
in Northern Ireland	= 1½ units
in Scotland	= 1¼ units

supplied by foods, although the amounts required are very small.
3 A deficiency of any vitamin will cause a specific disease, failure to grow normally in children and a general feeling of ill health.

Historically many of the *deficiency diseases* have been known for centuries, and so have some of the foods that will cure them, but it was not until the turn of this century that the vitamins were isolated from foods and identified. This knowledge plus improvements in the standard of living has meant that deficiency diseases are now rare in the UK and it is only those who choose from a narrow range of foods or who are already ill that are susceptible to them. However, in the Third World vitamin deficiencies are more common and are associated with poverty and food shortage.

Vitamins were originally named by letter in order of their discovery and then given chemical names as more detail became available. They can be subdivided into two groups. First, vitamins A, D, E and K which are soluble in fat and not in water and hence are found in fatty foods. It is possible to take in an excess of any vitamin by eating large amounts of fortified foods and megadoses of vitamin pills. This very occasionally causes toxic effects from the fat soluble vitamins A and D as the excess cannot be removed in the urine but builds up in fatty tissues. In the second group are the B vitamins and vitamin C which are soluble in water. These vitamins can be lost from food during cooking if the cooking liquid is discarded and are also destroyed by heat. The instructions on pages 113 and 114 are aimed at keeping these losses to a minimum.

The functions of the vitamins and their main food sources are given in Table 1.8. This table shows the importance of fruit and vegetables in providing vitamins as well as supplying energy and fibre. It is also notable that a diet containing some red meat and dairy products will provide ample A and B vitamins. *Vegans*, who avoid all foods from animal sources, have more difficulty consuming sufficient of these vitamins, but it is

Table 1.8 *The main sources and functions of the vitamins*

Vitamin	Function	Sources
Fat soluble Vitamin A	Essential for vision in poor light and for healthy skin and surface tissues	Liver (33%), kidney, eggs, dairy produce (18%), fortified margarine (10%)
Carotene, the plant source of vitamin A	Essential for vision in poor light and for healthy skin and surface tissues	Carrots, dark green vegetables, tomatoes, total vegetables (21%), apricots, melon
Vitamin D	Aids *calcium* absorption from food and controls calcium levels in the blood and bones	Sunlight, fish liver oils, fatty fish, fortified margarine (45%), eggs, liver, fortified cereals (10%)
Vitamin E	Possibly needed for fertility	Vegetable oils, cereals, eggs
Vitamin K	Essential for blood clotting	Widespread in vegetables
Water soluble Thiamin or B_1	Controls the release of energy from carbohydrate and alcohol	Wholegrain cereals, offal, pork, eggs, milk, vegetables and fruit, total cereals (50%)
Riboflavin or B_2	Essential for the release of energy from food	Milk (29%), cheeəe, offal, beef, eggs, wheatgerm, fortified cereals, Marmite
Niacin	Essential for the release of energy from food	Offal, meat, fish, cheese, pulses, wholemeal bread, potatoes, coffee. The amino acid tryptophan can also be converted to niacin
Vitamin B_6	Involved in amino acid formation	Meat, fish, eggs, wholegrain cereals, peanuts, bananas, potatoes
Vitamin B_{12}	Essential for blood cell formation in conjunction with folic acid	Offal, meat, fish, dairy produce, eggs, fortified cereals
Folic acid or folate	Prevents some forms of *anaemia*	Small quantities in many foods, especially offal and raw green vegetables (35%); easily destroyed in cooking
Vitamin C	Necessary for the maintenance of *connective tissue* and wound healing	Blackcurrants, oranges, fruit juices, total fruits (40%), green peppers, green vegetables, potatoes (22%)

The percentage figures in brackets indicate the average contribution of that food to the vitamin intake, in the UK. Source: *The National Food Survey for 1984* (MAFF, 1986).

possible to achieve the correct balance, as explained in the section on balancing the diet (see page 31). Vegans, however, usually need to take supplements of vitamin B₁₂ as the only plant foods where B₁₂ is found are soya products and some seaweeds.

There are several exceptions to these general comments and the following vitamins require special mention.

Vitamin D can be made by the body but only with the aid of direct sunlight on the surface of the skin. However, as many people keep well covered or have little chance to go into the sunshine some dietary vitamin D is recommended. The natural sources of vitamin D are few and in 1984 45% of the intake in the UK came from fortified margarine. The other foods high in vitamin D are oily fish and liver, both of which tend to be unpopular, and deficiency of this vitamin does still occur amongst adolescents and the elderly, especially Asians in the northern parts of the country.

Folic acid is the vitamin most often deficient in the UK diet. Deficiency is most likely to occur during pregnancy as requirements are doubled and body stores of this vitamin are small. There is also evidence that low folate intakes in the very early days of pregnancy are associated with *spina bifida* which is a spinal defect that affects newborn babies. Folic acid deficiency is also likely to occur in old age: a recent survey of elderly people showed that 14% of those over age 70 had folic acid deficiency. Folic acid is a vitamin which is found in vegetables, but is very easily destroyed by heat, and therefore the deficiency is more common amongst those who unwittingly overcook their vegetables or who rely on snack foods with few fresh vegetables rather than traditional meals.

Thiamin or vitamin B₁ is needed to release the energy from carbohydrates and alcohol and the body can only store about one month's supply. Thiamin deficiency occurs in alcoholics who have a high requirement and often low intakes of many nutrients; it has also been noted amongst hunger strikers.

Minerals

The minerals are chemicals found in rocks and the soil and widely distributed in foods. They are required by the body for bone structure, for body regulation, or as an essential component of body fluids.

Two minerals, sodium and iron, are particularly important for health. The remainder, although essential, are found in such a wide variety of foods that they rarely cause problems.

Sodium
Sodium, together with chlorine, forms common **salt**, a substance widely used for food preservation and food production. Both sodium and

chlorine are essential for body fluid regulation and sodium is also required for muscle and nerve activity. However, the amount of sodium required is very small and it is estimated that it could be provided by 1 g of salt per day. The average intake of salt in the UK is between 8 and 12 g per day and this salt is the major source of sodium, as fresh or unprocessed foods contain very small amounts.

High salt intakes are associated with **hypertension** or high blood pressure, which is a major risk factor for coronary heart disease, and there have been a series of recommendations worldwide to reduce salt levels. In 1981 the WHO Expert Committee on Prevention of Coronary Heart Disease stated that people should not eat more than 6 g of salt per day. The NACNE report recommended an average reduction of 3 g per head per day and the COMA report said that our intake of salt was needlessly high and ways should be found of reducing the quantity of salt already in food as purchased. All the expert committees agree that large reductions in salt intake will not be easy to achieve quickly.

It is difficult to reduce salt intake as approximately 70% of salt consumed is added to food during manufacture. Bread, cereal products and meat products such as bacon, ham and sausages are the main sources of added salt in the UK diet. Some of this salt is unnecessary and food manufacturers are beginning to reduce levels in items such as bread, biscuits and tinned vegetables. However, salt is a preservative as well as a flavour enhancer and if salt is reduced too much, the storage life of food may be affected. The remaining 30% of salt is added during cooking or at the table, often from habit rather than necessity.

The use of salt in cooking in a catering establishment varies with the chef and is rarely measured. In addition, accurate measurement is difficult because of the small quantities involved, but there are many ways in which a caterer can monitor and reduce the use of salt. The following suggestions should be used in conjunction with Table 1.9 (see overleaf), which gives the relative sodium content of foods.

1 **Menu planning** Check the number of salted meats and fish used and always offer a fresh alternative.
2 **Measure** the salt used in a recipe, there are small battery operated digital scales now available to do this. The aim should be 0.5 g per portion for savoury dishes.
3 **Stocks** should be fresh whenever possible as the commercial bouillons and stock cubes are very high in sodium; alternatively make instant stocks up to half strength and avoid further seasoning in the dish.
4 **Seasoning** can be achieved using spices and herbs as an alternative to salt.

Table 1.9 *The sodium content of some foods*

High sodium foods (over 1000 mg/100 g)	Medium sodium foods (999–151 mg/100 g)	Low sodium foods (150 mg/100 g or less)
Bacon and ham	Burgers	Fruit of all kinds,
Black pudding	Cornish pasties	Fresh meat and
Luncheon meats,	Frankfurters	poultry
chopped pork	Tinned meats	Vegetables, fresh and
and salamis	Pork pies	frozen
Sausages	Cheddar, Edam and	Oils, all types
	cottage type	Tea and coffee
Blue cheeses	cheeses, butter and	Fresh milk and yogurt
Processed cheeses	margarine	
		Fresh white and oily
Smoked fish		fish
Prawns and shrimps	Kippers, crab, lobster,	Eggs, unsalted nuts
	scampi	Pasta and rice
Instant potato	Tinned fish and fish	Unsalted butter, cream
powder	fingers	Sugar and preserves
All-Bran		Fruit juices and
Cornflakes, Rice	Tinned vegetables	squashes
Krispies	Bread, cakes and	
	biscuits	
Instant stock,	Custard powder,	
Tomato Ketchup	self-raising flour,	
Piccalilli, Bovril,	Weetabix, chocolate	
Marmite, salt,		
Baking powder	Mayonnaise, salad	
	cream, tomato	
	sauce, tomato	
	purrée, instant	
	soups and sauces	

5 **Soups and sauces** will always be much lower in sodium if made on the premises rather than made up from tins or dehydrated products.
6 **Service** Never pre-salt foods prior to service, for example chips. The customers should choose how much salt they require.
7 **Staff training** Explain the danger to health of excessive or random use of salt to all staff responsible for cooking and serving foods.

It is obviously important to achieve a balance between highly salted dishes and those which are too bland and prove unpopular or encourage the customer to use excessive table salt. A change in policy is best implemented slowly as people are known to adapt well to gradual reductions in salt but have objected to a sudden cutback.

Finally, it is important to note that salt is not the only concentrated source of sodium in foods, as it is also a constituent of many raising agents, additives and preservatives. The following list gives some of the common ones and their uses: sodium bicarbonate, baking powder, raising agents; sodium nitrite, preservative and colour enhancer; monosodium glutamate, flavour enhancer; sodium benzoate, preservative; sodium alginate, emulsifier; sodium sulphite, preservative.

Potassium

Potassium is a mineral found in the fluid of the cells and it has a lowering effect on blood pressure. Therefore it is an advantage to eat foods with proportionally more potassium than sodium. Fruit and fruit juices are particularly useful in this context, but potassium is also found in vegetables, meat and milk.

Iron

Iron is an essential component of the red blood cells which carry oxygen around the body. These cells have a life of only 3–4 months, but the iron that they contain is retained and re-used in the formation of new red blood cells. A reserve supply of iron is also stored in the liver and this is necessary as it is so poorly absorbed from foods.

The most useful sources of iron are meat and offal because 25% of the iron in these foods can be absorbed. Other food sources are eggs, vegetables, bread and fortified cereals; but they yield only about 5% of their iron. Similarly, the iron in vitamin pills is poorly utilized. High vitamin C intakes increase iron absorption and the tannin in tea reduces it.

Losses of iron from the body occur whenever bleeding takes place and increased requirements are needed for growth. Thus, it is women of child-bearing age and children who are most likely to suffer from iron deficiency anaemia, a disease which is deceptively slow in onset but causes increasing tiredness and lack of vitality.

Iron deficiency is the most common nutrient deficiency in the world and is thought to affect 500 million people. The UK is no exception, although it is not a problem for the general population; the DHSS recommendations for iron intakes show that women need far more iron than men.

Children age 0–6 years require 6–8 mg iron per day
Children age 7–8 years require 10 mg iron per day
Children age 9–17 years require 12 mg iron per day
Men require 10 mg iron per day
Women age 18–54 years require 12 mg iron per day
Pregnant women require 13 mg iron per day
Lactating women require 15 mg iron per day

These last figures cover 90% of women and young girls, however anyone with heavy menstrual losses may need more iron.

Table 1.10 gives a summary of the functions and main sources of the minerals. The only two which are included in the DHSS table of recommended daily amounts of nutrients are iron and calcium (see Appendix 1).

Table 1.10   *The main functions and sources of the minerals*

Mineral	Function	Sources
Iron	Essential for the transport of oxygen in the blood	Meat and offal (21%); bread and cereals (42%); potatoes and other vegetables (17%)
Calcium	Essential for the structure of bones and teeth, and the function of muscles and nerves	Milk, milk products and cheese (54%); bread and cereals (25%); green vegetables. Calcium cannot be absorbed without Vitamin D
Phosphorus	Combines with calcium to form strong bones and teeth	Present in nearly all foods
Sodium	Regulates body fluids outside the cells. Essential for muscle and nerve activity	Salt, cured and processed foods. Intake exceeds requirement
Potassium	Regulates body fluids inside the cells	Vegetables, meat, fruit and fruit juices
Zinc	A component of more than 50 enzymes	Meat, wholegrain cereals and legumes
Iodine	A component of the hormone thyroxine which regulates basal metabolic rate	Seafood, iodized table salt
Fluorine	Traces of fluorine protect children's teeth from decay	Added to drinking water in most areas of the UK

The percentage figures in brackets indicate the average contribution of the food to the intake of that mineral in the UK.
Source: *The National Food Survey for 1984* (MAFF, 1986).

Energy and nutrient balance

A balanced diet has frequently been defined in terms of the total amounts of energy or particular nutrients needed to maintain health. However, it is now realized that this is a very narrow view which obscures the importance of the proportions of nutrients supplying the energy. Health can only be maintained if a balance is achieved by maintaining the following.

1 Appropriate total energy intake.
2 The correct proportion of that energy being derived from carbohydrate, protein and fat.
3 The diet providing sufficient vitamins, minerals and dietary fibre for individual needs.
4 Avoidance of an excessive intake of any one nutrient.

Total energy intake

The need for energy is dependent on a combination of factors, namely age, size, sex, growth requirements and occupation (see Table 1.1).

The DHSS has issued general guidelines on amounts of energy and nutrients required by groups of the population and these are given in Appendix 1. These guidelines have to be based on average needs and are not amounts that an individual must eat, as anyone can differ from the average. Also energy needs are not static as the body can adapt to changes in requirement or intake. For example, repeated attempts to diet often lead to a person needing less and less food to stay at a stable weight as the body adjusts to the lower energy intake.

The easiest method of assessing the accuracy of energy intake in adults is to monitor body weight. If this is gradually increasing, then energy intake is too high and vice versa. However, it is possible to have a stable weight but still be overweight or obese as energy intake may have been too high in the past. The most accurate measure of acceptable weight is given by a combination of weight and height as illustrated in Figure 1.2, which is reprinted from the Health Education Council's *Guide to Healthy Eating*. This gives a series of weight bands ranging from underweight to the very fat or obese and the individual's band is found by plotting height without shoes against weight. In 1983 the report on obesity by the Royal College of Physicians stated that over one-third of adults are in the overweight to obese range. Excess weight can be removed by a combination of eating less food energy, as suggested in Figure 1.2, and increasing exercise. However, anyone who is obese should seek medical help on how best to achieve weight loss.

ARE YOU A HEALTHY WEIGHT?

Take a straight line across from your height (without shoes) and a line up from your weight (without clothes). Put a mark where the two lines meet.

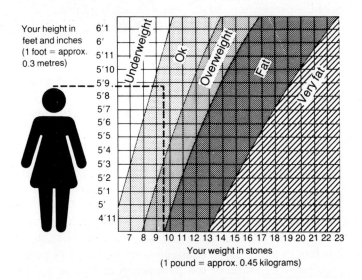

Your height in feet and inches (1 foot = approx. 0.3 metres)

Your weight in stones
(1 pound = approx. 0.45 kilograms)

UNDERWEIGHT Maybe you need to eat a bit more. But go for well-balanced nutritious foods and don't just fill up on fatty and sugary foods. If you are *very* underweight, see your doctor about it.

OK You're eating the right *quantity* of food but you need to be sure that you're getting a healthy *balance* in your diet.

OVERWEIGHT You should try to lose weight.

FAT You need to lose weight.

VERY FAT You urgently need to lose weight. You would do well to see your doctor, who might refer you to a dietitian.

If you need to lose weight

Aim to lose 1 or 2 pounds a week until you get down to the 'OK' range. Go for fibre-rich foods and cut down on fat, sugar and alcohol. You'll need to take regular exercise too.

Figure 1.2 *Chart to show ideal body weight. Adapted from the Health Education Council, London, with kind permission. The dividing line between each weight band is a guide rather than a precise measurement.*

Children up to the age of 17 years require relatively more energy than adults because they are growing and also they tend to be more active. Appetite should control their energy intake, as any unecessary restrictions can lead to stunted growth or disturbed eating habits such as anorexia nervosa. Visibly overweight children require medical advice to find the cause of their obesity. This is often a high intake of sweets and high-fat snack foods and can also be associated with lack of exercise.

Nutrients and energy

Carbohydrate, fat, protein and alcohol all provide energy and it has been shown that the current national eating pattern causes ill health because too much of that energy comes from fats, sugar and alcohol and too little from unrefined carbohydrates. Balancing the proportion of energy derived from each nutrient type is just as important as achieving the correct total energy intake. The long-term dietary goals recommended by the NACNE report are compared with average intakes in Figure 1.3. If alcohol is not included in the total energy calculations, the figure for recommended total fat comes out a little higher at 35% of energy.

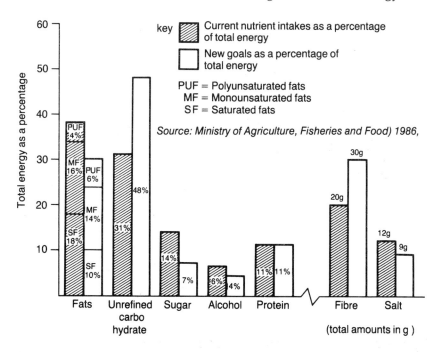

Figure 1.3 *A comparison of current nutrient intakes and the long-term NACNE recommendations.*

The percentage of energy provided by each nutrient within an average daily intake, a meal or a recipe is calculated as follows.

1 *Carbohydrate*

Total carbohydrate in grams $\times$ 16 = kJ of energy from carbohydrate

$$\frac{\text{kJ of energy from carbohydrate} \times 100}{\text{Total energy}} = \% \text{ of energy from carbohydrate}$$

2 *Fat*

Total fat in grams $\times$ 37 = kJ of energy from fat

$$\frac{\text{kJ of energy from fat} \times 100}{\text{Total energy}} = \% \text{ of energy from fat}$$

3 *Protein*

Total protein in grams $\times$ 17 = kJ of energy from protein

$$\frac{\text{kJ of energy from protein} \times 100}{\text{Total energy}} = \% \text{ of energy from protein}$$

4 *Alcohol*

Total alcohol in grams $\times$ 29 = kJ of energy from alcohol

$$\frac{\text{kJ of energy from alcohol} \times 100}{\text{Total energy}} = \% \text{ of energy from alcohol}$$

The calculation can be done using kcal by substituting the following factors.

Carbohydrate	multiply by 4 instead of 16
Fat	multiply by 9 instead of 37
Protein	multiply by 4 instead of 17
Alcohol	multiply by 7 instead of 29

The result in kcal is slightly less accurate than using kJ.

Example 1

A 30-year-old man has an energy intake of approximately 12,000 kJ per day and eats a daily average of 97 g fat, 413 g carbohydrate and 77 g protein. He also consumes 17 g of alcohol per day.

$$\frac{16 \times 413 \times 100}{12,000} = 55\% \text{ of energy from carbohydrate}$$

$$\frac{37 \times 97 \times 100}{12,000} = 30\% \text{ of energy from fat}$$

$$\frac{17 \times 77 \times 100}{12,000} = 11\% \text{ of energy from protein}$$

$$\frac{29 \times 17 \times 100}{12,000} = 4\% \text{ of energy from alcohol}$$

TOTAL   100%

Example 2
The man's wife, who is of the same age and moderately active, has an energy intake of 9000 kJ per day. She eats a daily average of 97 g fat, 255 g carbohydrate, and 65 g protein, but does not often drink alcohol so that her average intake is 5 g per day.

$$\frac{16 \times 255 \times 100}{9000} = 45\% \text{ of energy from carbohydrate}$$

$$\frac{37 \times 97 \times 100}{9000} = 41\% \text{ of energy from fat}$$

$$\frac{17 \times 65 \times 100}{9000} = 12\% \text{ of energy from protein}$$

$$\frac{29 \times 5 \times 100}{9000} = 2\% \text{ of energy from alcohol}$$

The two examples have the same total fat content but with different consequences as the fat is eaten as part of food intake with varied energy levels. Example 1 meets the dietary goals but example 2 has too much energy coming from fat and too little from carbohydrate. This shows the importance of knowing both the *total* amount of a nutrient, for example fat, and the *percentage of the total energy* which that nutrient provides when assessing the quality of a food or a diet.

The provision of vitamins, minerals and fibre

The foods that are available to choose from are numerous but fall into six general groups.

Group 1   Cereals, bread, pasta, rice, potatoes and root vegetables.
Group 2   Fruit and all other vegetables.
Group 3   Meat, poultry, fish, eggs and the vegetable protein alternatives, nuts and pulses.
Group 4   Milk and milk products.

If some foods from each of these four groups are used daily when menu planning the resultant diet is unlikely to be lacking in any of the vitamins or minerals. In addition, the foods in Group 1 will provide the bulk of the dietary fibre needed if a high proportion of wholegrain cereals are used and carbohydrate is providing more than half the total energy.

The animal foods in Groups 3 and 4 are high in saturated fatty acids, therefore the lower fat alternatives, chicken, white fish, skimmed milk and lower fat cheeses, should be chosen regularly but not to the exclusion of lean red meat. For instance, a 7-day main meal cycle could include meat three or four times, chicken one or two times, fish one or two times and a vegetable quiche or similar item once.

Group 5   Sugar, cakes, biscuits, sweets and sweet drinks.
Group 6   Fats and oils.

These two groups contain the high-energy dense foods and some essential fatty acids. Except for the very active, their use should be limited.

Excessive intakes of a nutrient

Any diet where one food or nutrient is predominant is also likely to contain a narrow range of other foods and therefore be unbalanced. In particular, those who derive a high proportion of energy from alcohol can lack the B vitamins and protein. Similarly high sugar and high fat diets are often associated with low complex carbohydrate intakes and therefore lack dietary fibre.

Salt and megadoses of the fat soluble vitamins A and D are the individual nutrients to avoid in excess.

The key to a balanced nutrient intake is to choose from as wide a variety of foods as possible and the majority of the population do use a mixture of all the foods from Groups 1–4.

Vegetarians, however, avoid some or all animal products but nevertheless have lower rates of obesity, coronary heart disease, hypertension and colon disease than the general population. Lacto-vegetarians obtain adequate amounts of protein from milk products and eggs instead of animal flesh. *Vegans*, who avoid all animal products, need to use a wide variety of pulses, nuts and cereals as protein sources to obtain a balanced mix of amino acids and they also need to supplement their diets with vitamin B$_{12}$. Also vegan diets are bulky and special care needs to be taken to ensure that vegan infants obtain sufficient energy and calcium.

Summary

This chapter has detailed the links between food and health and the areas of concern in current eating patterns. Figure 1.3 shows the changes needed as percentages of total energy. However, it is also useful to know how much of an individual nutrient the dietary goals represent. This is less easy to define as individual people are so varied in their size and activities. The recommended amounts of food energy and nutrients given in Appendix 2 give minimum levels for specific groups of the population but do not include figures for carbohydrate, fat, salt or fibre. In 1986 the British Medical Association in its document *Nutrition, Diet and Health* gave figures for the national average diet based on the long-term NACNE proposals and this included figures for fats, carbohydrates and salt. Table 1.11 is a summary of these proposals which will act as a guide, but will obviously need to be adapted for individual needs.

Table 1.11 *Average daily diet*

Energy	Sufficient to maintain optimal body weight and adequate exercise	
Protein	56 g	11% of energy*
Total fat	72 g	30% of energy
Saturated fatty acids	27 g	10% of energy
Sugar	56 g (2 oz)	A reduction of 50%
Fibre	30 g	An increase of 50%
Unrefined carbohydrates	No limit	Likely to be 50% of energy
Alcohol for men	3 units†	A reduction of 10%
Alcohol for women	1½ units†	A reduction of 10%
Salt	9 g	A reduction of 3 g

*Protein figure based on the average energy intake in 1984 of 8700 kJ.
†Units of alcohol are defined on page 20.

The comparison between average eating habits and those being recommended by the Government and the Medical Profession shows that large changes in the intake of many foods are needed. Changes of this size cannot be achieved quickly and it is envisaged that it could take up to 15 years to implement the long-term proposals. For this reason the NACNE report also suggests a more modest set of dietary goals to be achieved in the 1980s.

In terms of energy the new total diet should consist of

Protein	11% (no change)
Fat	34% (a reduction)
Carbohydrate	50% (an increase)
Alcohol	5% (a reduction)

Health relies upon an informed and balanced choice of food throughout the week. Although no one meal can be described as nutritionally right or wrong, the more often a meal high in fat or sugar is eaten, the harder it becomes to achieve an overall balance.

The current trends in eating show that on average people take more than three meals per week outside the home. This is obviously an advantage to the Catering Profession but also means that caterers have an important role to play in the chain of food availability and they too must **cater for health**. It will become damaging for the industry if food choice in the home is seen to be healthy and food choice outside is invariably high in *fat, sugar* and *salt*. The most frequently served meals in 1986 certainly did not compare favourably with the NACNE guidelines. Analysis of the most popular dinners show fats as a percentage of energy ranging from 55% to 62%.

The two essential ways to improve the nutrient balance offered by the Catering Industry are first to choose the raw materials with care and second to use cooking methods and recipes that do not add unnecessary fat, sugar or salt, and therefore maintain nutrient balance. Chapter 2 gives guidance on the former and the recipe section on the latter.

Further reading

Textbooks

Manual of Nutrition. 1985. Ministry of Agriculture, Fisheries and Food. HMSO.
Human Nutrition and Dietetics. 1986. R. Passmore and M. A. Eastwood. Churchill Livingstone.
Nutrition Guidelines. 1985. ILEA. Heinemann Education.

Government and medical reports and documents.
Eating for Health. 1978. DHSS. HMSO.
McCance and Widdowson's—The composition of foods. 1978. A. A. Paul and D. A. T. Southgate. 4th edition. HMSO.
Recommended daily amounts of food energy and nutrients for groups of people in the United Kingdom. 1979. *DHSS Report on Health and Social Subjects No 15.* HMSO.
The NACNE Report. Proposals for nutritional guidelines for health education in Britain. 1983. NACNE. Health Education Council.
The COMA Report. Diet and Cardiovascular Disease. 1984. *DHSS Report on Health and Social Subjects No. 28.* HMSO.
Nutrition, Diet and Health. 1986. The British Medical Association.

Pamphlets

Eating for a Healthier Heart. The British Nutrition Foundation and the Health Education Council.
Healthy Eating. Health Education Council.

CHAPTER 2

Food commodities

Introduction

Food purchasing is one of the most important functions in any catering operation and it becomes of even greater importance when nutrient balance is considered as well as the provision of acceptable meals. It is essential to know the main nutritional features of a recipe ingredient before it is chosen and the following details on food commodities should be used in conjunction with the recipe section of this book.

There are three main sources of information for the nutrient content of food.

Food tables, which give average figures for the composition of food. These figures are derived from the analysis of several randomly sampled batches of a similar food as purchased. When using food tables, care should be taken to check the description of the food as there are many variations, particularly with meat and milk products.

Food manufacturers and distributors, many of whom carry out detailed analysis of their own products.

The Food Labelling Regulations, which are discussed in detail below.

The Food Labelling Regulations

The Food Labelling Regulations, which were amended in 1984, give some indication of food content. The regulations cover foods to be delivered as such to the ultimate consumer or to caterers, with these exemptions:

Milk, which is covered by separate regulations.
Food not intended for human consumption.
Food intended for export.
Food for consumption by Her Majesty's forces.

The regulations state that food should be marked with a list of ingredients in descending order of weight determined 'as at the time of their use in the preparation of the food'. This ingredient list gives the purchaser a guide to the raw materials used in processed and prepared foods and the additives that they contain. However, the following foods do not have an ingredient list.

Fresh fruit and vegetables
Carbonated water and vinegar
Cheese, butter, fermented milk and fermented cream
Flavourings
Any food consisting of a single ingredient (flour is included in this category)
Any drink with an alcoholic strength by volume of more than 1.2%

Thus the purchase of raw, fresh foods is in general exempt from any detailed labelling.

The ingredient lists give no indication of the total amounts of food or nutrients but only an indication of the predominant raw materials. In response to public demand, many retail outlets are now providing a nutrient analysis on their packaging, but this type of information is rarely found on catering packs. These nutrient analyses are not always easy to use as there is no uniform system for presenting the information, which may be given per 100 g, per ounce, per pack, or per portion.

Also some of the information can be misleading because it does not include all the relevant nutrition information needed for the food, but only selected items. Two recent publications aimed at the caterer and the general public illustrate this problem.

'The Caterers guide to dairy products', issued by the Milk Marketing Board, includes a section on healthy eating which fails to include the need to reduce saturated fatty acid intake and does not include fat type in its nutrient analysis tables. However 61% of the fatty acids in milk are saturated, which is the highest level found in any common food (see page 11).

'Eggs in action', issued by the Egg Authority, has a section on nutrition which neglects to point out that eggs are one of the few common foods which contain cholesterol (see page 45).

In the light of current interest in nutrition labelling, the MAFF are preparing a set of guidelines aimed to ensure that nutrition information given voluntarily is laid out in a standard format. At the time of writing the draft guidelines suggest three progressively more detailed formats.

All nutrients must be specified per 100 g or per 100 ml in a specific order.

Format (a)
Fat, protein, carbohydrate, energy, in that order. In accordance with the Fat Content of Food (Labelling) Regulations, fat must be broken down to show saturates.

Format (b)
As (a) with carbohydrates broken down to show sugars separately. In addition, sodium should be listed after the energy content.

Format (c)
As (a) and (b) followed in alphabetical order by the vitamins and minerals present in the food.

Additional information can include dietary fibre, polyunsaturates, monounsaturates and starch, but no other nutrient type.

These proposals have aroused much discussion, particularly as they do not include provision to itemize added sugar and also because many nutritionists would prefer the use of high, medium, and low symbols as well as numeric information.

The most practical method a food purchaser can use to obtain full and unbiased nutrition information on products is to draw up a checklist of the details required. The manufacturer's packaging and literature can then be checked against the list and any gaps in the information identified. Further details are often available either from the manufacturer, food tables or from a dietitian. This is a particularly important procedure for bulk purchases or standing orders of prepared food items where there is a wide range of products to choose from. The checklist could include the following:

Fats	Total fat content
	% energy from fat
	Saturates
	Polyunsaturates
	% fat and lean in meat products
Protein	Total protein
Carbohydrates	Total carbohydrate
	% energy from carbohydrate
	Natural sugars
	Added sugars—sucrose and glucose
	Fibre
Energy	Kilojoules and/or kilocalories
Sodium	Total sodium—to compare with 3.6 g (3600 mg) per person per day
	Salt—to compare with 9 g per person per day
Vitamins	If required
Minerals	If required

A further source of nutrition information will be the proposed legislation on the labelling of the fat content of food which was recommended by the COMA report.

The Fat Content of Food (Labelling) Regulations 1987 will be laid before parliament during 1987 and, providing they become law, the two most important sections of the regulations will take effect immediately, namely:

Regulation 1   No person shall sell any food unless it is marked or labelled with a declaration as to its actual fat content. Foods with a variable fat content such as fresh meat may be labelled with a typical fat content.

Regulation 2   No person shall sell any food unless it is marked or labelled with a declaration as to its actual saturated fatty acid (SFA) content. Foods may also be marked with a typical SFA content where appropriate.

The remainder of the regulations describe the units to be used and suggest a system of declaring fat content by food category rather than individual variety; for example plaice and sole would be labelled as white fish. However, these sections were still under discussion at the time of going to press.

Foods to be exempted from the Regulations are:

1   Any food or drink containing less than 0.5% fat other than milk in returnable glass bottles.
2   Fresh fruit and vegetables. .
3   Herbs.
4   Spices.
5   Seasonings.
6   Alcoholic drinks containing more than 1.2% alcohol.
7   Bread and flour as defined in the Bread and Flour Regulations 1984 (a).
8   Cereals as defined in the Bread and Flour Regulations 1984.
9   Food sold for immediate consumption in or from any catering establishments, with the exception of those products which are subject to Regulation 8.

Thus, the catering sector in general will not have to label its products, but Regulation 8 is aimed at fast food chains which 'produce a limited range of products with a standard formulation intended for consumption outside the establishment'. This regulation will be a great advantage to the consumer, as fast food meals tend to have a very high level of their total energy coming from fat. Although several organizations supply nutrition leaflets to their customers, the information given rarely contains details of fat type. The emphasis given to nutrient information can also be misleading. The leaflet issued by McDonalds describes their French fries as a valuable source of vitamin C rather than a food in which 48% of the energy comes from fat. Wimpy describe a meal of tea, fish and chips as having an acceptable level of fat when fat provides 44% of the energy in that meal. The information supplied by Wimpy in their Fact Sheet No. 1 is particularly confusing as it compares the fat content of meals to an average day's intake rather than the energy content of the meal itself.

The information on food commodities in this chapter has been compiled by using all the available sources of information.

Cereals

Cereal foods provide energy as they are high in starch and contain significant amounts of protein; they are also good sources of the B vitamins, particularly thiamine, and minerals. Cereals are the main source of protein for vegans and contribute approximately 20% of the average protein intake in the UK. Potentially cereals are a rich source of **dietary fibre** but how much fibre they contain varies with the plant source and extent to which they have been refined during manufacture.

Flours

Wheat flour is the most commonly used flour for cooking. It is manufactured by grinding and sifting whole wheat grains and the amount of outer husk which is removed will determine both the type of flour and its fibre content. The outer husk or bran is an excellent source of high quality fibre, as is **wholemeal flour** which contains 100% of the original wheat seed. However, the amount of bran in **white flour** (plain flour) is very small as only the interior of the grain is retained. Wholemeal flour can be substituted for white in many recipes but its colour and coarser texture make it less suitable for sauces and sweet items. The bran in wholemeal flour absorbs moisture from other recipe ingredients slowly and this means that the amount of liquid used needs to be increased and the mixing technique adapted to prevent hard or dry products.

All types of wheat flour are difficult to wet and will form lumps when added to liquids. It is for this reason that they are usually dispersed in fat, to form a roux, when used as thickening agents. However, fat is not the only suitable ingredient in which to mix flour, tomato purée and any of the skimmed milk cheeses, such as fromage frais, are equally suitable. Otherwise the technique of the slow addition of liquid combined with vigorous mixing, as used in making batters, can be used. This is the principle employed for the thickening paste in the recipes.

Besan flour is produced from dehulled and ground chick-peas and is widely used in Indian cookery. The fibre content is high (approximately 7%) and it makes an excellent substitute for plain white flour in most sauces, cakes and pastries. Besan flour looks yellow when raw and gives off a distinctive smell of peas when mixed with liquids, but both these properties diminish during cooking. The thickening power of besan is slightly greater than that of wheat flour and the starch cooks out more quickly. Besan flour is now widely available in the UK, much of it being imported from Australia, and is sometimes sold under the generic name of **gram flour**.

Cornflour, extracted from maize, and **arrowroot** are both flours which are almost entirely starch. They have a fine particle size which makes them easy to mix with liquids and thicken readily at boiling point.

Breads

The nutrient composition of bread reflects the type of flour used in baking. There is detailed legislation covering the minimum composition of breads but the variety of names used for loaves is confusing.

Wholemeal or **wholewheat** bread contains all of the wheat grain.

Granary breads contain added malted grains but are not necessarily made entirely with wholemeal flour, and some have sugar added to improve the crust colour.

Brown bread is made from flour which must have a minimum of 6% fibre (dry weight). Caramel is often added to improve the colour.

White bread is made from highly refined flour with a low fibre content. The Bread and Flour Regulations ensure that the vitamins and minerals lost during milling are replaced as additives. White bread also contains bleaching agents and preservatives to prevent mould growth.

High fibre white bread or bran enriched white bread has bran added, but this is not necessarily wheat bran because of its brown colour. The source of the bran is often pea husks.

All breads have salt added during manufacture to control yeast action, to retain moisture in the loaf after baking, and to improve the flavour. Salt is essential but the levels are often unnecessarily high, especially in mass production.

Pasta

Until recently all pasta was made with high-protein, white flour and in some varieties additional ingredients such as eggs or spinach. Wholemeal pastas are now available with an increased fibre content but with very similar cooking properties to traditional pastas. All pastas are good sources of starch and energy.

Rice

White rice has the outer husk removed during milling and may be par-boiled (easy cook) to reduce the cooking time and prevent the cooked grains from sticking. **Brown** rice has not been dehusked and contains almost double the amount of fibre to white rice. Brown rice has a pleasant nutty flavour but needs fast boiling for approximately 30 minutes to soften the husk. Catering size packs of 'easy cook' brown rice are available and this variety cooks in 20 minutes and is suitable for rizottos.

Breakfast cereals

The fibre, starch and sugar content of these products is very variable (see page 9). Many breakfast cereals are high in fibre and make a valuable contribution to fibre intakes. These brands are based on wholewheat and usually give their fibre content on the packaging, but it is the amount of fibre per portion which is the important figure to check. Table 2.1 gives some indication of the fibre content of the different types of cereals; those with 2.5 g of fibre or more per portion have been marked with ***. The fibre content of cornflakes is not necessarily as high as the figure given in the food tables and those brands with a sugar coating should be avoided.

Table 2.1 *Guide to the fibre content of cereals*

White bread	**	All Bran	****
Brown bread	***	Bran Flakes	****
Granary bread	***	Cornflakes	*
Wholemeal bread	****	Grapenuts	***
		Muesli	***
Besan flour	****	Puffed Wheat	***
Cornflour	*	Porridge	**
Plain flour	**	Rice Crispies	*
Wholemeal flour	****	Shredded Wheat	***
		Weetabix	***
Brown rice	***		
White rice	**		
Spaghetti	**		
Wholemeal pasta	****		

Rich sources of fibre are marked ****
Moderate sources of fibre are marked ***
Poor sources of fibre are marked **
Negligible sources of fibre are marked *

Milk, cream and cheeses

Milk

Milk is one of the most complete foods eaten as it contains protein, carbohydrate, fat and a wide range of vitamins and minerals. The exact amount of these nutrients depends on the animal source of the milk and how the animal was fed. For example the fat content of cow's milk ranges from 3.8–4.8%, goat's milk contains 4.5%, human milk 4%, and ewes milk has the lowest level at 3%.

The fat in all animal milks is high in saturated fatty acids therefore any product made with whole milk or milk fat will also contain a high proportion of saturated fatty acids.

Liquid whole milk contributed 17% of the total saturated fatty acid intake in the UK during 1984. This was the largest contribution of any single food, but, it is also the easiest to reduce as there are several lower fat alternatives to whole milk available (Table 2.2).

Table 2.2   *Fat content of different types of milk*

Milk type	Bottle top code	Fat content (%)
Whole milk		
Channel Islands	Gold	4.8
Pasturized	Silver	3.8
Semi-skimmed	Red striped	1.5–1.8
Skimmed	Blue checked	<0.3
Dried skimmed	–	As fresh

Note. Dried skimmed milk powder with added vegetable fat has a variable saturated fatty acid content, many products are based on palm and coconut oil.

The remaining nutrients in the different milks occur in similar amounts, with the exception of vitamin A, which is progressively removed with the fat.

Packaging
Domestic sales are traditionally based on household deliveries of 1 pint returnable glass bottles, but milk is now increasingly purchased in 1, 2 or 4 pint cartons.

Catering sales   Fresh whole milk is sold in catering packs of 3 or 5 gallon capacity but fresh skimmed milk can only be purchased by the pint.

Individual portions (14 ml) of skimmed and fresh UHT 'Long Life' are available.

The use of reconstituted dried skimmed milk reduces the fat content of a dish without any effect on product quality and has the advantages of easy storage and lower cost.

Specialist milks such as Vitapint (J. Sainsbury Plc) and Shape (St Ivel Foods) are based on skimmed milk with added non-fat milk solids to improve the texture. Nutritionally they are lower in fat than whole milk and slightly higher in protein and lactose, the sugar of milk.

Yogurt

Yogurt is produced by adding a culture of lactic acid bacteria to milk under controlled conditions. The nutrient content of yogurts varies according to the milk base used and any additional ingredients used.

Natural low-fat yogurts are similar to skimmed milk, although some have added skimmed milk powder which gives them a higher protein content. These blend well with cream, desserts and mayonnaise but curdle if brought close to boiling point.

Flavoured yogurts are either whole or skimmed milk based and contain considerable amounts of sugar.

Greek yogurt is made from boiled milk and contains 12% fat.

Cream

Cream is the fat of whole milk with a variable amount of whey and is a much more concentrated source of saturated fatty acids than milk, therefore it should be used sparingly. The precise description and fat content of the different creams are covered by the Cream Regulations 1970 (Table 2.3).

Table 2.3 *Fat content of different types of cream*

Cream type	Minimum fat content (%)
Half cream	12
Single cream	18
Whipping cream	35
Double cream	48
Clotted cream	55

The main catering uses for cream can all be adjusted to use a type with a lower fat content or an alternative product, for example low-fat yogurt or fromage frais. When cream is to be used as a topping or piped garnish, whipping cream properly whisked will trap an almost equivalent volume of air (overrun). This is an important point when serving cream as a sauce or topping (see page 217).

Cheese

Cheese has the image of a high protein and high calcium food but many cheeses are also very high in fat and salt. There are over 400 cheeses commonly available made from cows', goats' and sheep's milk. The cheeses commonly used in cookery are listed in Table 2.4.

Table 2.4 *Cheeses used in cookery*

Cheese	Fat content (%)	Comments
Cream	47	Use alternatives, cottage or quark.
Cheddar	33	Mustard enhances the flavour and enables a lower recipe weight to be used. Use a low-fat hard cheese alternative.
Cottage	4	Liquidize to produce a smooth alternative to cream cheese.
Fromage frais	1–8	A soft cheese of French origin used to make salad dressings, dips and sauces (see pages 164–170). Not acidic and goes well with fresh fruit. Blend with flour as thickening agent. Use in place of cream to lighten sauces and desserts.
Fromage blanc	1–8	The same as fromage frais.
Quark	1	A skimmed milk, soft cheese of German origin similar to fromage frais but thicker in consistency, which makes it an ideal ingredient when firmer mixtures are required (see cheesecake on page 223).
Bodyline (Express Catering Foods)	14.5	Low-fat hard cheese similar to cheddar and available in 2.5 kg cater packs. Can be used as a substitute for cheddar in most recipes but should not be subjected to prolonged heating.
Parmesan	30	Expensive and high in fat—use sparingly.

The cheese board

The choice of cheeses can reflect fat content as well as flavour and texture:

High fat	Stilton
	English hard pressed cheeses
	Danish blue
	Goat's milk cheeses
Moderate fat	Gouda and Edam
	Brie and Camembert
Lower fat	Feta
	Cottage

Serve cheese with brown bread, brown rolls or low-fat biscuits such as matzos or water biscuits. Also accompany with a selection of prepared raw vegetables, for example celery sticks, radishes, strips of carrot, and red and green peppers.

Eggs

Eggs are a versatile commodity invaluable in cookery because of their capacity to foam and act as a binding agent. The white contains protein and minerals and the yolk is a good source of protein, iron and vitamin D. However, egg yolk is also one of the few dietary sources of cholesterol and therefore eggs should not be eaten in excess of three or four per week, especially by people known to be at risk from heart disease.

Fats and oils

Fats and oils contain a high proportion of fat (between 81 and 99.9% of their weight), water and some fat soluble vitamins. The only exception is the low-fat spreads which contain more than 50% water bound to the fat with emulsifying agents and stabilizers. The most important nutritional feature of a fat is the mix of fatty acids that it contains (see page 11).

Table 2.5 gives a score for the proportion of fatty acids typically found in different fats and oils. The precise composition of many products, particularly the blended oils, varies with market fluctuations in the price of the raw materials used. Caterers should check for any changes when ordering a new batch of cooking oil. The introduction of the Fat Labelling Regulations will make this a much easier task (see page 38). Unfortunately the oils which are high in polyunsaturates are also amongst the most expensive, but as demand increases the economics of scale should lower the price.

Table 2.5 *Guide to the proportion of fatty acids typically found in different fats and oils*

Fat/oil type	Fatty acid (FA) profile PFA	SFA	Comments
Spreading fats			
Butter	*	†††	Spread thinly, replace with polyunsatured margarine in cooking, offer alternatives at the table.
Margarine (hard)	*	††	Contains trans-fatty acids (TFA), which are equivalent to SFAs. Replace with poly-unsaturates where possible.
Soft margarines: vegetable oil base	***	†	High in polyunsaturates, creams well and will increase the volume of cakes. Check the label.
Soft margarines: vegetable/animal oil	**	††	Variable in composition, may also contain TFAs. Check before ordering.
Sunflower margarine	***	†	Excellent FA profile. Available in 10 g portions, 4 boxes containing 100 × 10 g per case, and 250 g packs.
Low-fat spread	**	††	Variable in composition. Separates on heating. Most suitable for use on weight-reducing diets.
Salad oils			
Palm oil	*	†††	Avoid using this oil if possible. Also note its regular use in oil blends.
Groundnut oil	**	††	Intermediate FA mix. Frequently used in oil blends.
Olive oil	*	†	High in neutral monounsaturated fatty acids. Use in combination with PSF oils to improve flavour.
Corn oil	***	†	Use for salads, brush roasting and shallow frying in place of butter or dripping.
Rapeseed oil	**	–	A recent introduction with a very low level of SFAs, increasingly used as the major constituent of vegetable oil blends.
Soya bean oil	***	†	Use for salads, brush roasting and shallow frying in place of butter or dripping.
Sunflower oil	***	†	Use for salads, brush roasting and shallow frying in place of butter or dripping. Has good shortening properties in pastries.
Proprietary blends	*	†††	Variable in content. Check for the use of coconut oil which is very high in SFAs.

Table 2.5 *cont.*

Fat/oil type	Fatty acid (FA) profile PFA SFA		Comments
Safflower oil	***	†	Very high in PSF. Use for salads, brush roasting and shallow frying in place of butter or dripping.
Cooking fats			
Lard	*	†††	Replace with polyunsatured oil or margarine.
Dripping	–	†††	Avoid keeping this.
White margarine	***	†	Produced for the domestic market for pastry making, frying and roasting. Based on sunflower oil with hydrogenated vegetable oils, so will contain some TFAs. Use instead of lard. Available in 250 g packs.
Pastry margarine	*	†††	Replace with polyunsatured oil or margarine.
Frying oils			
Solid fat base	*	†††	Poor keeping properties and high uptake by foods. Replace with long-life brand.
Oil base (low price range)	*	†††	Variable content, poorer coating effect and therefore higher uptake by foods.
High-life brands	*	†††	Better coating properties which seal the food before too much oil is absorbed and make it easier to shake off the excess.
High-life polyunsatured oil	***	†	Best choice but see notes on frying, page 141.

Key to Table 2.5

Fat category	Polyunsaturated fatty acids (PFA)	Saturated fatty acids (SFA)
High	***	†††
Moderate	**	††
Low	*	†

Meat, poultry and fish

Meat, poultry and fish are the major sources of protein and because they are eaten in relatively large amounts in the UK, they also supply approximately 20% of energy. All are sources of B vitamins and iron and, as well, liver provides 33% of vitamin A and oily fish 14% of vitamin D. However, the amount and type of fat in these three protein foods differs.

Meat

The fat of meat is high in saturated fatty acids and it is important to choose carcases and cuts which have both a light fat covering and a low *lean muscle* fat content.

The proportion of fat in meat is very variable and depends on several factors.

Animal type   Sheep have the highest levels, followed by pigs, and cows the lowest levels.

Breed   Some breeds are naturally leaner.

Age   Younger animals have laid down less fat.

Feed   Grass- and silage-fed cattle have less saturated fats than grain-fed cattle.

Caterers and butchers can buy leaner carcases if they specify their requirements. The Meat and Livestock Commission (MLC) operate a classification scheme which specifies fatness.

Beef
There are seven fat classes:

Very lean						Very fat
1	2	3	4L*	4H	5L	5H

*Light = light; H = heavy.

Currently a typical beef carcase is in fat class 4L. A leaner carcase with a lower fat content can be achieved by specifying fat class 2 or 3.

Lamb and mutton
There are five fat classes:

Very lean				Very fat
1	2	3	4	5

Fat class 3 may be divided into 3L and 3H and a typical sheep carcase is in this class.

Table 2.6 *A guide to the grade symbols for New Zealand export sheep meats*

Lamb grades

Symbol	Weight range	Wrap print colour	Fat content
PL	9.0–12.5 kg 19.5–27.5 lbs	Blue	Medium
PM	13.0–16.0 kg 28.5–35.5 lbs	Blue	Medium
PX	16.5–20.0 kg 36.5–44.0 lbs	Blue	Medium
PH	20.5–25.5 kg 45.0–56.0 lbs	Blue	Medium
YL	9.0–12.5 kg 19.5–27.5 lbs	Red	Light
YM	13.0–16.0 kg 28.5–35.5 lbs	Red	Light

Lamb carcases are allocated to one of six grades on the basis of fat content and weight.

Mutton & Hogget grades

Symbol	Weight range	Wrap print colour	Fat content
Hogget grades			
HX	22.0 kg (48.5 lbs) and under 22.5 kg (49.5 lbs) and over	Red	Light
HL	22.0 kg (48.5 lbs) and under 22.5 kg (49.5 lbs) and over	Blue	Medium
Mutton grades			
MM	All weights	Black	Almost devoid
MX	22.0 kg (48.5 lbs) and under 22.5 kg (49.5 lbs) and over	Red	Light
ML	22.0 kg (48.5 lbs) and under 22.5 kg (49.5 lbs) and over	Blue	Medium
MH	All weights	Blue	Heavy
MF	All weights	Black	Excessive

HOGGET GRADES. Hogget carcases are allocated to one of two grades on the basis of fat content.
MUTTON GRADES (EWES & WETHERS). Mutton carcases are allocated to one of five grades on the basis of fat content.

The New Zealand Meat Producers Board introduced a stricter grading scheme in 1984 to encourage leaner meat production. It contains two new Y grades for lamb with a light fat cover and one each for mutton and hogget. These are shown in Table 2.6 (see page 49).

Pork

The Meat and Livestock Commission grade the fat content of pork according to a measurement taken on the long loin above the last rib bone. The depth of fat is measured in millimetres and is indicated on the carcase. The lower the measurement, the less fat the carcase (Figure 2.1).

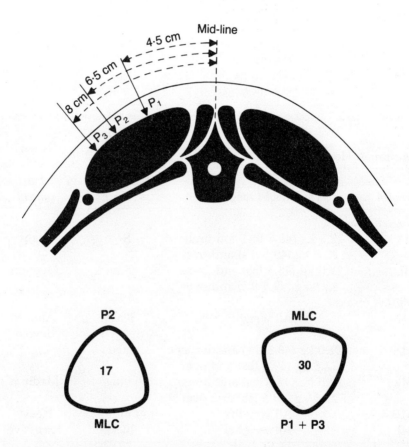

Figure 2.1 *Measuring and marking in grading the fat content of pork. A probe is inserted at fixed points P_1, P_2 or P_3, level with the head of the last rib. The two main methods of measurement are explained in this text. (See page 51.)*

There are two main methods of measurement:

Method 1 A single measurement is taken at P_2. This indicates depth of fat over the loin. Stamp on the belly or fore leg.

Method 2 Measurements taken at P_1 and P_3. Average fat depth over the loin is half the total on stamp. Stamp on hind trotter.

Purchasing lower fat class animals not only reduces the external or visible fat on meat, it also reduces the amount of hidden fat in the lean of meat. For example:

	% fat in the lean eye muscle
Pork loin with medium fat cover	7.8
Pork loin with light fat cover	5.7

Alternative cutting methods for lamb and pork
The Meat and Livestock Commission have developed alternative cutting methods for lamb and pork which are designed to produce a range of cuts and joints to suit modern taste. The methods, which involve boning out the animal carcase and seaming out individual muscles, produce leaner cuts and joints which are attractive and easy to cook and carve. Lower fat carcases are usually used for these cutting techniques and the joints and cuts can be well trimmed of fat.

The National Association of Catering Butchers (NACB) adopts the specifications for butcher meats laid down in *The Meat Buyers' Guide for Caterers*, written by three founder members of the NACB with assistance from the Meat and Livestock Commission. The text includes details on a wide range of joints and cuts and highlights important specifications on fat content. Terms are used to denote the various cuts—'larder trim' indicating a well trimmed product. All the specifications on the cuts and joints have a computer compatible code for ease of ordering from suppliers.

Meat purchasing

Bulk purchasing by caterers

1 Carcases: specify the fat class.
    (a) beef, lamb and mutton: 2 or 3 using the MLC grades.
    (b) New Zealand lamb: YL or YM.
    (c) mutton: MX or MM.
    (d) Hogget: HX.
    (e) pork: P_2 fat depth: 12 mm (carcase weight 55 kg), or P_2 fat depth: 15 mm (carcase weight 75 kg).

2 Cuts and joints: specify depth of surface fat—8 mm maximum (reduced from standard 13 mm).
3 Diced meat: specify visual fat content—not more than 10%.
4 Mince: specify fat content—not more than 10%.

Small quantity purchasing
The amount of visible fat on meat is a good guide to total fat content, but this is often partly obscured by the packaging used in a supermarket. Purchasing from a good butcher avoids this problem and a butcher will also know the grade of carcase. Extra lean cuts are now increasingly available and, although more expensive, avoid the waste and cost involved in trimming fat during preparation. Table 2.7 gives a quick reference guide to the relative fat content of common cuts of red meat, poultry and game. Regular use of the many varieties in the low-fat categories will reduce saturated fat intakes.

Poultry and game

Chicken and **turkey** meat are both lower in total fat and saturated fatty acids than red meat and as such make good menu alternatives that should be included at least twice in any 7-day main meal cycle. **Duck** is higher in fat than chicken and turkey, but the majority of this is associated with the skin. If the breast meat only is used it becomes similar to chicken in fat content. In contrast both the flesh and skin of **goose** are high in fat.

Game birds are wild rather than reared and therefore do not build up fat stores until older and tough, so it is best to choose young roasters and not casserole birds. The level of polyunsaturates in game birds is often high, with **grouse** the first bird in season also the first on the list.

Polyunsaturated fatty acids in poultry and game *(% of total fatty acids)*	
Grouse	61
Rabbit	34
Turkey	29
Partridge	25
Chicken	16
Duck	12
Pheasant	12

Table 2.7 *The average fat content of common cuts of raw meat, poultry and game*

High-fat meats (fat content 31–40% of raw weight)	*Low-fat meats* (fat content 5–10% of raw weight)
Duck—including skin	Pheasant
Streaky bacon	Ox liver
Back bacon	Partridge
Best end of lamb	Calves liver
Belly of pork	Turkey—meat and skin
Loin of lamb	Pigs liver
	Venison
Moderately high-fat meats (fat content 21–30% of raw weight)	Chicken liver
Loin of pork	Lambs hearts
Scrag and middle neck of lamb	*Very low-fat meats* (fat content less than 5% of raw weight)
Shoulder of lamb	
Goose	Duck—flesh only
Fore rib of beef	Chicken—flesh only
Sirloin steak—lean and fat	Grouse—flesh only
Leg of pork	Rabbit
	Lambs kidney
Moderate-fat meats (fat content 11–20% of raw weight)	Pigs kidney
Leg of lamb	Veal—lean
Chicken—including skin	Ox kidney
Ox tongue	Turkey—flesh only
Minced beef	
Rump steak—lean and fat	
Pigeon	
Topside of beef	
Stewing steak	
Lambs liver	

Note. The meats are arranged in descending order of fat content within each category. The source of information is *McCance and Widdowson's Composition of Foods* and will not include any changes in the fat content of meats brought about by the use of leaner breeds or new butchery techniques.

Fish

White fish contain very little fat and **oily** fish have moderate amounts, but these are mainly polyunsaturated and include the essential fatty acids. All fish can be included in menus regularly and customers encouraged to choose fish dishes by the use of imaginative preparation.

Meat products

Meat products are often very high in fat. In 1984 sausages and meat products accounted for 9.5% of total fat and 9.2% of saturated fatty acids eaten in the UK.

The content of sausages and meat products is controlled by Food Regulations (1967/68) which state a minimum meat content. Table 2.8 gives some examples.

Table 2.8 *Fat and meat content of various meat products*

Product	Minimum meat content (%)	Average fat content (%)
Frankfurter, Vienna sausage	75	25
Salami	75	45
Pork sausage	65	32
Meat with jelly	80	23
Meat with gravy	75	13
Luncheon meat	80	27
Faggots	35	18
Pork pie	25	27

These, often high, meat levels, as illustrated in Table 2.8, should leave little room for fat, rusk, water or any other ingredients. However, the definition of meat used in these regulations is: *the flesh, including fat, and the skin, rind, gristle and sinew in amounts naturally associated with the flesh of any animal or bird normaly used for human consumption.* Thus, the meat itself can contain substantial amounts of fat, as shown by the average fat content of the products listed above.

An increasing range of reduced fat sausages are now available produced by brand leaders such as Walls (their new pork sausage contains 13% fat) and own brand food stores (Sainsbury's lower fat sausages contain 8% fat). The new Fat Content of Food (Labelling) Regulations will inform purchasers of the fat levels in meat products, but these foods are also high in salt and other sodium-containing additives. Therefore detailed product specifications should be sought on items such as pies, pasties and ready-to-serve meals.

Pulses and nuts

Pulses are the seeds of the legumes—peas, beans and lentils—and nuts, the seeds of a variety of trees. Both pulses and nuts are particularly good sources of dietary fibre; second only to wholewheat. They also provide protein, some fat, vitamins and minerals. Pulses are particularly useful as meat extenders or meat alternatives because of their fibre and protein content. The recipes which contain pulses have both a higher fibre content and lower percentage energy from fat than those with just vegetables added.

Pulses contain unusual simple carbohydrates that may not be digested in the normal way by people unused to eating them. Instead, they are broken down in the large intestine by bacteria and gases are liberated. However, the digestive system soon adapts to the presence of these carbohydrates.

There is a wide range of pulses available, as shown in Table 2.9 (see overleaf). The soya bean is the basis for two commonly used meat substitutes.

Tofu

This is a cheese-like product made from fermented soya bean curd. It has a fairly high protein content of 7%, and has a fat content of 4% (mainly polyunsaturated fatty acids). However, it is low in fibre.

Textured vegetable protein (TVP)

TVP is prepared from soya bean protein which is then spun or extruded to produce meat-like shapes. TVP contains approximately 50% protein, dry weight, and is fortified with the vitamins and minerals normally found in meat. It does not contain fat but will absorb fats more readily than meat when being cooked. TVP is most frequently used for vegetarian products or as a low-cost meat extender. A mixture of 10% (reconstituted weight) TVP mince and fresh mince gives a very acceptable end-product, however the TVP diced products blend less well with fresh meats.

Nuts are high in fat, but this is predominantly polyunsaturates, with the exception of coconut. They also contain protein, B vitamins and minerals and are useful alternatives to meat in vegetarian dishes.

Tinned pulse vegetables often contain added sugar and unnecessarily high levels of salt, so *check the label*; this is particularly noticeable with baked beans, red kidney beans and butter beans. Use fresh dried pulses whenever possible.

Table 2.9   *The range of pulses available and their cooking times*

Name of pulse (and description)	Cooking time (after soaking)	Country of origin	Weight increase from 50 g sample
Aduki Small reddish round bean; sometimes cooked with rice, tinting it pink.	50 minutes	China	110 g/120%
Black beans	15 minutes (P)*	Australia	140 g/180%
Black-eyed beans Quick-cooking bean with a pleasant flavour.	30 minutes	America	120 g/140%
Black kidney beans Large shiny bean; may be used as a substitute for red kidney beans.	90 minutes	China	120 g/140%
Brown lentils Retains shape well when cooking.	15 minutes	Canada	130 g/160%
Butter beans Large, flat kidney-shaped beans which collapse easily when overcooked.	40 minutes	USA	120 g/140%
Cannelini beans Small white kidney beans which make a good substitute for haricot beans.	15 minutes (P)	France	120 g/140%
Chick-peas Small, round and golden brown in colour; retain shape well when cooking.	15 minutes (P)	Turkey	110 g/120%
English field beans Round brownish coloured beans with tough skins.	20 minutes	Britain	110 g/120%
Flageolet Pale green beans with a fresh flavour.	45 minutes	France	100 g/100%
Haricot beans Known as 'Navy beans' in USA—used as baked beans.	30 minutes	Canada	110 g/120%
Mung beans Often used as a sprouting bean; quick cooking.	35 minutes	Thailand	130 g/160%

Table 2.9 *cont.*

Name of pulse (and description)	Cooking time (after soaking)	Country of origin	Weight increase from 50 g sample
Pinto beans Speckled pink coloured beans, a variety of kidney beans.	12 minutes (P)	USA	110 g/120%
Red kidney beans Mealy textured beans.	15 minutes (P)	USA	110 g/120%
Soya beans Contain the highest protein content of all beans—long cooking time.	20 minutes (P) (minimum)	Canada	120 g/140%

* P denotes the use of a pressure steamer.

Fruit and vegetables

The main contribution of fruit and vegetables is to supply low-energy bulk to the diet, vitamins, iron and minerals. Green vegetables contain vitamin C, carotene, folic acid, iron and minerals. Also, starch is supplied from potatoes.

Fruits contain fructose and some fruits, such as apples, are also high in pectins (a type of fibre with good water-holding properties).

The amount of fibre in fruit and vegetables is extremely variable and is best illustrated by giving the quantity found in an average serving (Table 2.10; see overleaf).

Tinned vegetables often contain added sugar and unnecessarily high levels of salt, so *check the label*, particularly for peas and carrots. Sugar-free and salt-free varieties are becoming increasingly available on the domestic market due to consumer demand. Catering packs will follow suit if bulk purchasers also specify 'no added sugar or salt'.

Tinned fruits are usually prepared in heavy syrups of sugar or glucose, so *check the label* and use sugar-free varieties whenever possible.

Stock cubes and bouillons

Stock cubes and bouillons are extensively used in food preparation but their major ingredient is **salt** and their use makes many of the resultant dishes very high in sodium. Many products also contain other sodium salts, for example monosodium glutamate (MSG), and azo dyes, which

Table 2.10   *The fibre content per portion of some fruit and vegetables*

Fruit/vegetable	Fibre (g)
Lettuce	<0.5
Orange	1.0
Onions	1.3
1 apple	1.5
1 pear	1.7
Cauliflower	2.6
Potatoes (peeled)	2.6
Leeks	3.1
Blackberries	3.1
Carrots	3.6
Cabbage	4.2
Peas (frozen)	4.7
Potatoes (unpeeled)	5.2
Spinach	6.0

give a distinctive yellow tinge and repetitive flavour to dishes. Bouillon should be used at half strength to reduce these problems and should never be allowed to boil down during cooking.

Vegetarian stock pastes are also available and these often have lower sodium levels. For instance **Vecon** contains no added salt; its sodium content is 6.9 g/100 g derived from yeast and vegetable extracts used in its manufacture.

Commercially prepared sauces, pickles and flavouring mixtures

These rely heavily upon salt and other sodium salts for their flavour and many of them are also very high in sugar. The sugar content of these products becomes more important when they are eaten in relatively large amounts. For example, the following sauces are served in measurable amounts, especially on meals such as a Ploughman's Lunch, and have sugar as their second ingredient by weight: Branston Pickle, Tomato Ketchup, Sweet Piccalli, Fruit pickles and Mango chutney. Anyone who regularly eats tomato ketchup with his/her meals is, in effect, sprinkling unnecessary sugar on savoury items.

The use of 'in house' made sauces, pickles and relishes will both reduce the salt and sugar content of meals and be an attractive menu feature.

Commercially prepared products should be used sparingly.

Soya and shoyu sauces

These are seasoning sauces made from fermented soya bean curd.

Tamari

Tamari is similar to soya and shoyu sauces but stronger in flavour.

Tahini

Tahini is a paste made from sesame seeds and sesame oil and contains 55% fat which is very high in polyunsaturated fatty acids. It is a basic ingredient of hummus and as this is also based on chick-peas, it is a nutritious vegetarian paste (see page 168).

Herbs and spices

Herbs are aromatic green plants which contribute to the flavour of dishes and will reduce the need for heavy seasoning. They are used in such small amounts that they do not affect the nutritional content of food, with the exception of fresh parsley which may contribute some vitamin C. Fresh herbs have delicate flavours and make attractive garnishes. Dried herbs are stronger and should be used more sparingly and stored in airtight containers. Herbs should be purchased regularly in small amounts and not left unused in the kitchen or stored in paper bags, as stale herbs can have overpowering and unpleasant flavours.

Spices are the seeds of plants and therefore contain proteins, starch and fats as they are the food store for the developing seedling. The amounts used in cooking are generally very small, with the exception of thick curry sauces (see page 82). The important contribution which spices make to cookery is their flavour, which reduces the need to add salt. Many Indian cookery books suggest high levels of salt in their dishes based on the need to use the salt as a preservative which is unnecessary in this country. Spices must be stored in airtight containers to retain their fragrance.

Food additives

A wide range of additives is used in refined and processed foods. Additives act as colouring ingredients, flavour enhancers, stabilizers or preservatives. Many of them are essential to product quality and shelf-life and all are covered by Food Legislation. Those permitted by the EEC are listed as ingredients with an **E number**.

Food additives are of two sources.

1 Naturally occurring substances found in foods which are extracted for use as food additives.
2 Artificial additives manufactured as chemicals and added to foods.

A small minority of people are affected by some food additives but it is often difficult to establish which chemical is responsible for their symptoms. The most frequent problems are caused by colouring agents.

Caterers should be aware of this potential problem for customers and be in a position to advise on the content of dishes served.

There are several books which cover this subject in detail.

1 *Statutory Instruments 1984 No. 1305. Food.* The Food Labelling Regulations 1984. HMSO.
2 *Food Legislation of the UK.* D. J. Jukes. Butterworth and Co, 1984.
3 *E for Additives.* Maurice Hassen. Thorsons Publishers Ltd, 1984.
4 *Allergy? Think About Food* (2nd edition). Susan Lewis. Wisebuy Publications, 1986.

Further reading

The Meat Buyers' Guide for Caterers. 1983. R. Moore, J. Stone and H. Tattersall, in association with the Meat and Livestock Commission. International Thomson Publishing Ltd.
Alternative Lamb Cutting Method. Meat and Livestock Commission.
Alternative Pork Cutting Method. Meat and Livestock Commission.
Beef Carcase Classification for Meat Traders. Meat and Livestock Commission.
Pig Carcase Classification. Meat and Livestock Commission.
Sheep Carcase Classification. Meat and Livestock Commission.
A Guide to the Grade Symbols for New Zealand Export Sheep Meats. New Zealand Lamb Catering Advisory Service.
Professional Cookery, The Process Approach. 1985. D. R. Stevenson, Hutchinson Education.

CHAPTER 3

Introduction to the recipes

The recipes in this book are arranged according to the process used. Each chapter is concerned with a method of cooking or the general preparation for a group of dishes. This means that dishes are mostly located under an appropriate method of cooking and not simply by menu or commodity. For example, minted lamb kebabs, which are grilled, are to be found in the chapter devoted to grilling, whereas blanquette of lamb, which is a stew, is located in the chapter dealing with boiling and stewing. A recipe index which lists all the dishes in the book has been included for the quick location of dishes.

Presenting information in this manner reveals the practical relationships between dishes and highlights basic cookery knowledge. *It is the intention to present recipes together with nutrient details which the reader may produce or use as a guide to create many more dishes.*

Practical implications

Many of the recipes in this book have an increased vegetable content compared with traditional recipes, therefore consideration should be given to preparation time. The *use of food processors, liquidizers and powered equipment which cut a variety of vegetable shapes* will considerably reduce preparation time and avoid production procedures of a laborious and tedious nature.

Nutritional implications

The cookery processes have been modified to produce dishes with a nutrient combination which is in line with the NACNE guidelines. The modifications have been achieved by a combination of adapting the methods used to cook the food and careful choice of the ingredients.

The aim of the recipes is to provide the following nutrient balance.

Fats

Energy from fat
The recipes have been formulated so that fat contributes approximately

35% of the total energy in the dish. This level is a considerable reduction in fat when compared with many dishes produced by caterers. There are some recipes where it has not been possible to achieve 35% because of either the basic ingredients or the importance of fat to the texture of the dish. These recipes are flagged★★ and can be used in combination with lower fat foods to provide a balance. Where fats have been reduced, alternative ingredients such as complex carbohydrates have been added. It is not the intention of the authors to produce low calorie recipes or fat-free recipes; this is unnecessary for the general public.

The objective is to offer meals with modest amounts of fat and to avoid the excessive addition of fat to raw materials which in themselves are relatively low-fat foods. A prime example of this is the practice of adding butter to green vegetables prior to service.

Fat type
Ingredients high in saturated fatty acids have either been reduced in quantity or substituted for polyunsaturated products.

Each basic recipe gives the total fat content per portion and has a chart showing how much of the energy in the dish comes from fat.

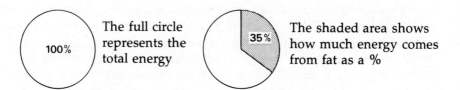

Where a method or the ingredients have been adapted from a traditional recipe to reduce the level of fat, the size of the reduction is shown by an arrow.

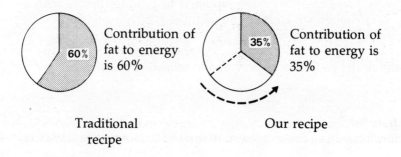

Carbohydrates

The reduction in fats should be matched with an increase in carbohydrates if energy levels are to be maintained and, in addition, there is a need to increase fibre intake. These two aims can be achieved simultaneously by including more wholegrain cereals, starches, vegetables and fruit in the diet. The recipes have been formulated to alter carbohydrate content as follows.

Fibre

The fibre content of the recipes has been increased in two main ways. First, by using wholemeal flour where it is in keeping with the end-product. Wholemeal flour, however, has a higher water-holding capacity than white flour and gives many dishes a less acceptable texture and colour. As an alternative chick-pea flour or **besan** has been used as this has a high fibre content of approximately 7 g per 100 g (see page 39) and produces cakes, pastries and sauces with a very stable texture.

Second, the fibre content has been increased by greater use of pulses, cereals and vegetables in the recipes. However, the most *significant increases* in fibre result when white bread is replaced by wholemeal and with the *increased* use of breakfast cereals.

The amount of fibre in the recipes is given as a total weight and also there is a chart to show how much fibre an average portion will contribute to the recommendation of 30 g per person per day.

30g | Full circle shows total fibre 30 g

1.4g | Shaded area shows fibre in one portion

Sugar

Sugar levels in many standard recipes are unnecessarily high and these have been reduced where possible. However, sugar is an essential ingredient in many bakery items, where it aids aeration and the development of colour. Sugar is also important in maintaining the texture of frozen goods and in the production of meringues. For these reasons the level of sugar has only been lowered slightly in some recipes.

Proteins

Animal protein foods frequently have associated fat, giving them a high saturated fatty acid profile. Therefore, lean cuts and grades of red meat have been used together with an average red meat portion size of 100 g raw weight. A selection of fish and poultry recipes are included as these are low in saturated fatty acids.

Vegetable protein foods have been used regularly either to augment meat or to provide vegetarian dishes.

Salt

The salt content of the recipes is based on 0.5 g per portion as an average. Salt levels can often be reduced below this for spiced dishes and those based on bacon or ham, but may need to be increased for bland foods such as fish.

The nutrient calculations

The data for the nutrient calculations in the recipes and the tables is based on *McCance and Widdowson's The Composition of Foods* by A. A. Paul and D. A. T. Southgate (HMSO, 1978), and the second supplement to this book *Immigrant Foods* by S. P. Tan, R. W. Wenlock and D. H. Buss (1985). The fibre content of besan flour was analysed independently. The dietary calculations were done by computer using the **Super Diet** dietary analysis program at the University of Surrey.

All the calculations and recipe development are based on metric weights and kilojoules. The final figures have been rounded up to the nearest whole number unless that number was less than 2, when one decimal place is included. Conversion to Imperial measure was achieved by multiplying the final metric weights in the recipes by the factor 28.35 then giving the weight in ounces to the nearest ½ ounce. This can never give completely exact conversions but is more satisfactory than using either a standard 25 g ounce or 30 g ounce, particularly when large quantities are involved as the rounding errors that occur become a significant factor in recipe yield.

The nutrient analyses **per portion** are based on the stated yield in the recipes and are designed to give a quick visual and numeric guide to the recipe content. Full details of the nutrients **per recipe** are given in Appendix 2 and these should be referred to if the number of portions from a given recipe is altered. The percentage analysis of energy from fat will not be altered by alterations to portion size.

Menu planning

Menu planning is usually done in two main ways:

1   A menu for a catering establishment where the customer chooses from
    a range of dishes but the caterer has no control over the total nutrient
    intake of the meal.
2   A menu planned on a weekly basis where all meals are offered and the
    caterer can ensure adequate nutrient balance.

In either case, the range of dishes selected is extremely important.
Caterers so often dictate an unhealthy balance to their customer by failing
to offer suitable alternatives to the popular but high fat, sugar or salt
foods. Whichever style of catering is involved there are several key points
to check.

Breakfasts

Always include:

*   wholemeal bread or toast
*   high fibre breakfast cereals, preferably low in sugar

This is the easiest way to increase fibre intake. For example, two slices of
wholemeal toast or two Weetabix supplies 5 g of fibre and that is one-
sixth of a day's requirement.

Always offer:

*   skimmed or semi-skimmed milk
*   polyunsaturated margarine
*   fruits; fresh or poached
*   natural yogurts and low-fat cheese, e.g., fromage frais

Cooked breakfasts should be poached, grilled or griddled rather than
fried.

Main meals

Starters
Use:

*   hors d'oeuvre which include fresh fruit or vegetables
*   low-fat soup recipes
*   fish dishes
*   rice and pasta dishes

Check the number of deep fried items and keep these dishes to a
minimum.

Main courses
Choose dishes based on poultry, fish and liver, kidneys, or other offal, as
often as those based on red meat. Always include a vegetarian dish.
    Check the style of cooking used:

- for the number of deep fried items because they supply between
  40–60% of their energy from fat.
- for the use of butter and cream in dishes and as a garnish.

    Use pastas, brown rice and jacket potatoes regularly as alternatives to
high-fat potato dishes, for example chips. These not only increase the
fibre content of a meal but also improve the nutrient balance, as the
following example of one portion of lamb kebab shows.

	% energy from fat	fibre (g)
Lamb kebab – unaccompanied	42	0
Lamb kebab with chips	47	3
Lamb kebab with tossed salad	48	<1
Lamb kebab with pilaf (see page 158)	28	3
Lamb kebab with pilaf and ratatouille	28	6

Offer lightly cooked vegetables with all meals as well as side salads which
contribute small amounts of fibre, but avoid serving with high-fat dressings.
    Avoid the steak, salad and chip syndrome as this type of meal is very
high in fat.

Desserts

Traditionally made desserts often contain unnecessary quantities of
cream and sugar. A combination of the recipes given in Chapter 9 will
give a range of desserts with an acceptable level of energy from fat. Any
selection of desserts should include fresh and light dishes to end the
meal, for example:

    fresh fruits
    sorbets
    low-fat ice-creams
    mousses and low-fat cheesecakes
    gateaux based on low-fat sponge and lightly decorated

Traditional favourites need not be heavy or high in fat, for example, fruit
flans and sponges, which can be served with a fruit sauce or low-fat
custard.

A random selection of recipes in the book will normally provide meals giving less than 35% energy from fat as the majority of the calculations do not include accompaniments such as potatoes and vegetables. Dishes which are higher in fat can be identified and balanced with the ones containing little fat.

Vegetarian meals

The following dishes from the recipe section of the book are suitable for vegetarians provided that they are made with vegetable stock or water.

Soups
Cauliflower soup or other thickened vegetable soup
Fenugreek soup
Gazpacho
Lentil soup
Tomato soup
Vichysoisse

Vegetable dishes
Channa dahl
Pasta with pine nuts
Ratatouille
Rice pilaff
Lentil dahl

Salads and spreads
Arabian salad
Baba Ganouge
Bean salad
Cheese and vegetable spread
Egyptian bread salad
Hummus
Moroccan carrot salad
Pasta salad
Turkish shepherd's salad

Main course dishes
Armenian vegetable stew
Macaroni and lentil bake
Pasta and potato bake
Pizza—tomato and vegetable

continued overleaf

Pizza—wholemeal tomato and vegetable
Potatoes filled with cheese and vegetables
Vegetable burgers
Vegetable curry
Vegetable dumplings
Vegetable hotpot
Vegetable hotpot with cheese
Vegetable lasagne
Vegetable quiche

Desserts
All the desserts in Chapter 9.

Gluten-free diets for coeliac disease

A person with coeliac disease has an allergy to the protein gluten, which is found in wheat, rye, barley, and oats. Coeliac disease is a relatively common problem, with an incidence in the UK of approximately one in every 2000 people. As a result, caterers are frequently asked to provide gluten-free dishes, and although this is not difficult, it can be boring for the customer, particularly if the desserts offered are always based on fresh fruit.

A gluten-free diet should not contain any wheat flour, which is present in the following foods: bread; most breakfast cereals; pastry; sausages and luncheon meat; biscuits and cakes; pasta; dishes thickened with a roux; commercial sauces and soups; most commercial ice-cream and dessert mixes; and proprietary sauces and ketchups.

Rye, oats and barley must also be avoided.

However, coeliacs can eat rice and vegetable foods including *besan flour*. Besan flour can be used as a substitute for wheat flour when thickening dishes and thus make them suitable for coeliac customers. Besan flour is also very successful when used in cake recipes as a substitute for flour as it gives products with a high volume and light texture.

Foods to use on a gluten-free diet are milk and yoghurts; fresh meat, poultry and game; cheese and eggs; margarine, butter and oils; breakfast cereals made from rice or maize; graveys made from cornflower or arrow-root; fresh fruit and nuts; and fresh vegetables and potatoes.

Further information and a current list of gluten-free manufactured products can be obtained from the Coeliac Society, PO Box 181, London NW2 2QY.

CHAPTER 4

Boiling and stewing

STOCKS

Good stocks are essential to the taste and flavour of sauces, soups and stews, particularly when very little or no *salt* is in the recipes.

Many different vegetables at hand may be used for preparing stocks, but never starch-containing vegetables such as potatoes, which would thicken and cloud the stock. Care must also be taken with vegetables which produce strong flavours, for example cabbage and brussel sprouts. These vegetables also produce off-flavours when cooked for prolonged periods, making them unsuitable for stocks.

Nutrient analysis is not given for these stocks because although they add flavour to dishes, they only contain very small amounts of soluble proteins such as gelatine. However, care should be taken if bacon or ham bones are used to make a stock as they will impart salt into the stock. The use of further salt in a subsequent dish will be unnecessary, and the stock itself should not be reduced down.

Cookery hints

- Ensure that all fat has been trimmed off the bones and any marrow removed from the centre of marrow bones.
- Cooking the bones in an oven, prior to making the stock, melts off any fat which is difficult to remove with a knife.
- Skim off all surface fat from stocks during cooking.
- Never allow a stock to boil rapidly as this emulsifies any fat droplets present and these will cloud the stock, and also will not float to the surface, so are difficult to remove.
- Trace fat on the surface of liquids can be easily removed by floating a piece of clean kitchen paper or dish paper on top—the paper will readily absorb the fat.
- When liquids are cold any fat which is present solidifies and can be easily removed from the surface.

Basic recipe for meat, poultry and game stocks

5 litres/8 ¾ pt

2 kg	Raw bones (beef veal, lamb, chicken)	4 lb 6 oz
300 g	Peeled onions	10½ oz
150 g	Peeled carrots	5 oz
150 g	Celery	5 oz
150 g	Leeks	5 oz
To taste	Thyme, bay leaf and parsley stalks	To taste
5 litres	Cold water	8¾ pt

Note: **Salty commodities** such as pieces of ham and bacon bones should be soaked in several changes of cold water to reduce the salt prior to use.

METHOD

1 Break up the bones and remove all *fat and marrow.*
2 Place the bones in a roasting tray and cook in an oven to melt off and remove any fat.
   White stocks: Do not allow the bones to develop any colour.
   Brown stocks: Allow colour to develop. Also, add the vegetables at this stage and allow to colour with the bones.
3 Place the bones and vegetables in a saucepan or stockpot then cover with the cold water.
4 Bring to the boil then slowly simmer for 6–8 hours.
   Note: Reduce the cooking time by half for poultry and feathered game stocks.
5 Skim off impurities and *surface fat regularly during cooking.*
6 Top up with additional water as and when required.
7 When cooked, strain into a clean container.
8 Remove any final surface fat when cold.

Fish stock

5 litres/8 ¾ pt

2 kg	White fish bones	4 lb 6 oz
400 g	Sliced onions	14 oz
To taste	Bay leaves and parsley stalks	To taste
60 ml	Lemon juice	2 large lemons
5 litres	Cold water	8¾ pt

METHOD

1   Place all the ingredients in a saucepan or stockpot and bring to the boil.
2   Skim off all impurities and surface fat and allow to simmer slowly for approximately 20 minutes. Also skim off all surface fat and impurities which develop during cooking.
3   When cooked, strain into a clean container.

Vegetable stock

5 litres/8 ¾ pt

Many different vegetables may be used when preparing this stock, except starchy vegetables (see beginning of section). Simply add vegetables to taste, but avoid using the seeds or pulp from vegetables such as peppers because a bitter taste may result. Preparing stocks is an ideal way of using vegetable trimmings.

Remember to cut most vegetables very small so that the maximum flavour is achieved in the shortest possible cooking time.

450 g	Carrot pieces (cut very small)	1 lb
450 g	Onion pieces (cut very small)	1 lb
220 g	Celery pieces (cut very small)	8 oz
220 g	Sliced leeks (cut very small)	8 oz
60 g	Mushroom trimmings	2 oz
5 g	Garlic	¼ oz
120 g	Celeriac (if available)	4 oz
300 g	Tomato trimmings	10½ oz
120 g	Asparagus trimmings (optional)	4 oz
60 g	Sweetcorn trimmings (optional)	2 oz
30 g	Watercress stalks (optional)	1 oz
60 g	*Chopped swedes	2 oz
60 g	*Chopped turnips	2 oz
60 g	*Cauliflower pieces	2 oz
60 g	*Sliced cabbage	2 oz
60 g	*Pieces of brussel sprouts	2 oz
To taste	Herbs: parsley, bay leaves, marjoram, thyme, etc.	To taste
5 litres	Water	8¾ pt

*These vegetables produce strong flavours and are also likely to develop off-flavours with prolonged cooking periods.

See overleaf for method

METHOD

1 Bring the water to the boil in a saucepan or stockpot.
2 Add all the vegetables and return to the boil.
3 Slowly simmer until the vegetables are cooked, 20–30 minutes.
4 Strain into a clean container.

Note: Strong vegetable stocks are made by increasing the quantity of vegetables in the recipe. Also additional flavourings such as tomato purée and commercial vegetable extracts (low salt) may be used to increase taste and flavour.

COOKERY HINTS FOR SAUCES, SOUPS AND STEWS

- The quantity of fat in recipes is often increased by many *traditional practices.* In some instances the increase in fat is considerable. For example, one method of preparing a tomato-type soup requires the recipe vegetables to be fried in fat and the soup liaised with cream, enriched with butter and garnished with cubes of bread, also fried in fat.
- Avoid shallow frying or sweating the recipe vegetables or garnishes for white sauces and white soups, for example white onion sauce and cauliflower soup. This is unnecessary and only increases the fat content.
- Use thickening pastes for sauces, soups and stews as an alternative to roux thickening. This means less fat is used and it is also a simpler means of thickening, saving time and labour.
- Use rich or fatty foods very sparingly. Many recipes often use unnecessary quantities of these foods. For example when producing gratinated dishes containing sauce and cheese, small amounts of grated cheese will achieve a satisfactory result; use plain sauce (not cheese sauce) with cheese sprinkled on top.
- Do not liaise sauces, soups or stews with cream unless it is necessary, for example cream soups; even then use as little cream as possible. See appropriate recipes for guidance on recommended quantities.
- Do not enrichen sauces, soups and stews with butter; a traditional practice (monter au beurre) for many dishes.

SAUCES

Nutrients for the sauces are given per 5 litre recipe in this section. Where a basic sauce is used as part of a later recipe, the nutrient analysis for the sauce is included with the other ingredients in that recipe.

The nutrient information for derivatives of the basic sauces, which will be used on there own, is given per 50 ml portion.

The use of all plain flour in these basic sauce recipes will reduce their fibre content by approximately half but will not appreciably alter the other nutrient levels.

White sauce (Béchamel)

5 litres/8¾ pt
80 portions

5 litres contain:
19099 kJ/4519 kcal,
24 g fibre, 180 g fat.

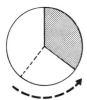

Energy
from fat 35%

4½ litres	Skimmed milk	7 pt 18 fl. oz
250 g	Onions—peeled and sliced	9 oz
2	Bay leaves	2
2	Cloves	2
200 g	Polyunsaturated margarine	7 oz
10 g	Salt	2 tspns
Thickening paste		
200 g	Besan flour	7 oz
200 g	Plain flour	7 oz
½ litre	Cold skimmed milk	18 fl.oz

METHOD

1 Place the milk, onion, bay leaves, cloves and margarine into a saucepan and bring to the boil.
2 Prepare the thickening paste by mixing together the two flours and cold milk to a smooth batter.
3 Whisk the thickening paste into the boiling milk and bring back to the boil.
4 Allow to simmer for 6–8 minutes then strain.

DERIVATIVES OF WHITE SAUCE

Anchovy sauce
Add anchovy essence to taste, 60 ml/2 fl. oz.

Mustard sauce
Add diluted English mustard to taste, 120 g/4 oz.

Onion sauce
Add 1 kg/2 lb 3 oz diced onions to the sauce when cooking. *Note:* Pierce and stud the bay leaves with the cloves into a large piece of onion for ease of removal.

Parsley sauce
Add chopped parsley until generously dispersed through the sauce.

Cheese sauce (Mornay)

5 litres/8 ¾ pt
100 portions

Each portion contains:
196 kJ/47 kcal, <1 g fibre, 1.5 g fat.

Energy
from fat 27%

4½ litres	Skimmed milk	7 pt 18 fl. oz
250 g	Onions—peeled and sliced	9 oz
2	Bay leaves	2
2	Cloves	2
80 g	Polyunsaturated margarine	3 oz
10 g	Salt	2 tspns
Thickening paste		
200 g	Besan flour	7 oz
200 g	Plain flour	7 oz
½ litre	Cold skimmed milk	18 fl. oz
400 g	Low-fat cheddar-type cheese	14 oz

METHOD

1 Place the milk, onion, bay leaves, cloves and margarine into a saucepan and bring to the boil.
2 Prepare the thickening paste by mixing together the two flours and cold milk to a smooth batter.

3 Whisk the thickening paste into the boiling milk and bring back to the boil.
4 Allow to simmer for 6–8 minutes then strain.
5 Add the cheese to the hot sauce on completion of cooking and stir until melted through the sauce.

Note: Do not reboil as the cheese may separate out of the sauce.

White sauces (veloutés)

(5 litres/8¾ pt)

5 litres contain:
8399 kJ/1993 kcal, 24 g fibre, 79 g fat.

Veloutés are white sauces which are made with stocks, therefore the flavour of the sauce will depend on the type of stock used.

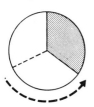

Energy
from fat 35%

METHOD

Prepare as white sauce on page 73, replacing the milk with the appropriate white stock, for example fish, chicken, veal or mutton. Also *reduce* the fat to 80 g/3 oz.

Brown sauce (sauce espagnole)

5 litres/8¾ pt

5 litres contain:
8971 kJ/2127 kcal, 24 g fibre, 79 g fat.

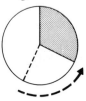

Energy
from fat 32%

80 ml	Polyunsaturated oil	3 fl. oz
4½ litres	Brown stock	7 pt 18 fl. oz
250 g	Sliced onions	9 oz
250 g	Diced carrots	9 oz
120 g	Sliced celery	4 oz
120 g	Sliced leeks	4 oz
	Sprig of thyme, 2 bay leaves and parsley stalks	
200 g	Tomato purée	7 oz
	Blackjack (gravy browning)	
10 g	Salt	2 tspns

Thickening paste

200 g	Besan flour	7 oz
200 g	Plain flour	7 oz
½ litre	Cold brown stock	18 fl. oz

METHOD

1 Shallow fry the vegetables in the oil until golden brown.
2 Add the tomato purée and stock and bring to the boil.
3 Allow to simmer until the vegetables are cooked, 45 minutes approximately.
4 Prepare the thickening paste by mixing the two flours and cold stock to a smooth batter.
5 Whisk the thickening paste into the boiling stock and vegetables and bring back to the boil.
6 Allow to simmer for 15 minutes then correct the colour with the blackjack. Strain and use as required.

DERIVATIVES OF BROWN SAUCE

5 litres/8¾ pt
100 portions

Barbecue sauce

Each portion contains:
117 kJ/28 kcal, 0.5 g fibre,
1 g fat.

200 ml	Vinegar	7 fl. oz
500 g	Chopped onions	18 oz
250 g	Chopped tinned tomatoes	9 oz
250 g	Chopped pickled onions	9 oz
250 g	Chopped gherkins	9 oz
120 g	Chopped capers	4 oz
60 g	Honey	2 oz
60 g	Diluted English mustard	2 oz
	Chopped parsley	
5 litres	Brown sauce	8¾ pt

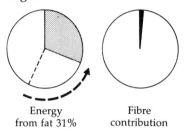

Energy
from fat 31%

Fibre
contribution

METHOD

1   Place the vinegar, onions, chopped tomatoes and honey into a sauce-pan and boil down by two-thirds.
2   Add the brown sauce, gherkins, capers, pickled onions and diluted mustard, bring to the boil and simmer for 4–5 minutes.
3   Add the parsley. Check seasoning and consistency.

Devil sauce

Each portion contains:
102 kJ/24 kcal, 0.5 g fibre,
<1 g fat.

500 g	Chopped onions	18 oz
500 ml	Vinegar	18 fl. oz
250 ml	White wine	9 fl. oz
To taste	Mignonette pepper	To taste
To taste	Cayenne pepper	To taste
5 litres	Brown sauce	8¾ pt

See overleaf for method

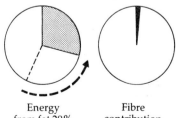

Energy
from fat 29%

Fibre
contribution

METHOD

1   Place the onions, mignonette pepper, vinegar and wine into a sauce-
    pan and boil down by two-thirds.
2   Add the brown sauce and simmer for 8–10 minutes.
3   Strain, then season with the cayenne pepper. Check consistency.

Brown onion sauce (sauce lyonnaise)

60 ml	Polyunsaturated oil	2 fl. oz
2 kg	Sliced onions	4 lb 6 oz
500 ml	Vinegar	18 fl. oz
120 ml	White wine	4 fl. oz
5 litres	Brown sauce	8¾ pt

Each portion contains:
152 kJ/36 kcal, 1 g fibre,
1.4 g fat.

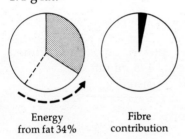

Energy   Fibre
from fat 34%   contribution

METHOD

1   Shallow fry the onions in the oil until golden brown.
2   Add the vinegar and white wine and boil down by two-thirds.
3   Add the brown sauce and simmer for 4–5 minutes.
4   Check consistency.

Piquant sauce

Each portion contains:
102 kJ/24 kcal, 0.5 g fibre,
<1 g fat.

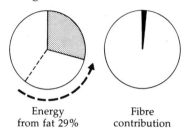

Energy
from fat 29%

Fibre
contribution

500 g	Chopped onions	18 oz
500 ml	Vinegar	18 fl. oz
120 ml	White wine	4 fl. oz
700 g	Chopped gherkins	1 lb 8½ oz
350 g	Chopped capers	12 oz
	Chopped parsley	
To taste	Cayenne pepper	To taste
5 litres	Brown sauce	8¾ pt

METHOD

1 Place the onions, vinegar and wine into a saucepan and boil down by two-thirds.
2 Add the brown sauce and simmer for 4–5 minutes.
3 Add the gherkins and capers and simmer for a further 2–3 minutes.
4 Add the chopped parsley and season with the cayenne pepper. Check consistency.

Bolonaise sauce

5 litres
40 portions
Each portion contains:
664 kJ/158 kcal, 1.4 g fibre,
6 g fat.

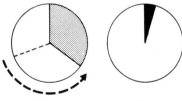

Energy
from fat 35%

Fibre
contribution

250 g	Red lentils, raw	9 oz
120 ml	Polyunsaturated oil	4 fl. oz
500 g	Chopped onions	18 oz
70 g	Crushed garlic	2½ oz
2 kg	Lean minced beef	4 lb 6 oz
140 g	Tomato purée	5 oz
2½ litres	Brown sauce	4 pt 8 fl.oz
10 g	Salt	2 tspns

METHOD

1 Cook the lentils in boiling water until soft, drain and reserve.
2 Shallow fry the onions and garlic in the oil until golden brown.
continued overleaf

3   Add the minced beef and continue shallow frying until lightly coloured.
4   Drain off any excess fat which has been produced.
5   Add the brown sauce and the lentils and bring to the boil. Check the consistency and if too thick add a little extra stock.
6   Simmer until cooked, 45 minutes approximately. Check consistency.

Bolonaise and vegetable sauce

40 portions

Each portion contains:
552 kJ/132 kcal, 2 g fibre, 5 g fat.

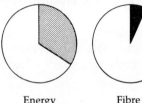

| Energy | Fibre |
| from fat 34% | contribution |

250 g	Red lentils, raw	9 oz
100 ml	Polyunsaturated oil	3½ fl. oz
500 g	Chopped onions	18 oz
70 g	Crushed garlic	2½ oz
170 g	Diced green peppers	6 oz
170 g	Diced red peppers	6 oz
170 g	Blanched diced carrots	6 oz
170 g	Blanched diced celery	6 oz
170 g	Diced courgettes	6 oz
350 g	Diced tomatoes (with seeds)	12 oz
1.3 kg	Lean minced beef	2 lb 14 oz
2¾ litres	Brown sauce	4 pt 17 fl. oz
10 g	Salt	2 tspns

METHOD

1   Cook the lentils in boiling water until soft, drain and reserve.
2   Shallow fry the onions and garlic in the oil until golden brown.
3   Add the minced beef and continue frying until lightly coloured.
4   Add the peppers, carrots, celery and courgettes and sweat for 4–5 minutes, then decant off any excess fat.
5   Add the tomatoes, lentils and brown sauce then simmer until cooked, 45 minutes approximately.

Note: The vegetables may be added near the end of the cooking period if a crisp texture is desired.

OTHER SAUCES
Tomato sauce

5 litres/8¾ pt

5 litres contain:
12,853 kJ/3049 kcal,
38 g fibre, 111 g fat.

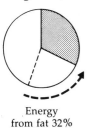

Energy
from fat 32%

120 ml	Polyunsaturated oil	4 fl. oz
4 litres	Bacon stock	7 pt
250 g	Onions	9 oz
250 g	Carrots	9 oz
120 g	Celery	4 oz
120 g	Leeks	4 oz
	Sprig of thyme, 2 bay leaves and parsley stalks	
1 kg	Tomato purée	2 lb 3 oz

Thickening paste

200 g	Besan flour	7 oz
200 g	Plain flour	7 oz
1 litre	Cold stock	1¾ pt

METHOD

Prepare the same as brown sauce omitting the blackjack.

Curry sauce (European style)

5 litres/8¾ pt

5 litres contain:
12,712 kJ/3027 kcal,
24 g fibre, 119 g fat.

Energy
from fat 35%

100 ml	Polyunsaturated oil	3½ fl. oz
3½ litres	Brown stock	6 pt
750 g	Chopped onions	1 lb 10 oz
30 g	Crushed garlic	1 oz
150 g	Curry powder	5 oz
120 g	Tomato purée	4 oz
250 g	Cooking apples (chopped)	9 oz
120 g	Mango chutney	4 oz
10 g	Salt	2 tspns

Thickening paste

340 g	Plain flour	12 oz
1 litre	Cold stock	1¾ pt

See overleaf for method

METHOD

1 Shallow fry the onions and garlic in the oil until soft, 10–15 minutes.
2 Add the curry powder and slowly cook for 3–4 minutes.
3 Add the tomato purée, cooking apples, chutney and brown stock and bring to the boil.
4 Simmer for 15 minutes then whisk in the thickening paste.
5 Allow to simmer for a further 15 minutes, then use as required.

Indian-style curry sauce

5 litres/8¾ pt

5 litres contain:
7239 kJ/1723 kcal, 20 g fibre, 63 g fat.

Energy
from fat 32%

30 ml	Polyunsaturated oil	1 fl. oz
1½ kg	Chopped onions	3 lb 4 oz
60 g	Crushed garlic	2 oz
120 g	Peeled root ginger	4 oz
60 g	Ground coriander	2 oz
30 g	Ground fennel	1 oz
60 g	Turmeric	2 oz
15 g	Cumin	½ oz
30 g	Cayenne pepper	1 oz
70 g	Garam masala	2½ oz
250 g	Tomato purée	9 oz
120 g	Fresh coriander (chopped leaves)	4 oz
500 ml	Low-fat natural yogurt	18 fl. oz
30 g	Lemon zest	1 oz
2½ litres	Chicken stock	4 pt 8 fl. oz

METHOD

1 Sweat the onions and garlic in the oil with the lid on until soft, 10–15 minutes.
2 Add all the spices and continue cooking for 6–8 minutes; avoid burning the spices.
3 Add the remaining ingredients, that is, tomato purée, chopped coriander leaves, yogurt, lemon zest and stock.
4 Bring to the boil and simmer for 30 minutes approximately.

Note: The sauce may be used as stated or liquidized to a finer consistency.

SOUPS

Soups offer the possibility of supplying a varied and interesting addition to a meal. Alternatively they may be a complete meal in their own right when heavily garnished and extended with vegetables, meat, poultry and game, etc. In addition, accompaniments such as wholemeal bread and bread rolls make the ideal partnership.

This section not only gives examples of how to prepare many traditional soups but should also provide guidance on how to create acceptable soups by way of fundamental procedures and recipes together with nutritional information.

PURÉE SOUPS
5 litres/8¾ pt, 25 portions

Lentil, bean and split pea soups

Each portion contains: 448 kJ/105 kcal, 5 g fibre, <1 g fat.

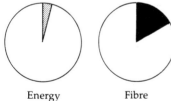

| Energy from fat 3% | Fibre contribution |

Basic recipe

5½ litres	Bacon stock	9 pt 12 fl. oz
250 g	Sliced onions	9 oz
250 g	Peeled carrots	9 oz
120 g	Sliced celery	4 oz
120 g	Sliced leeks	4 oz
	Sprig of thyme, 2 bay leaves and parsley stalks	
750 g	Pulse vegetable— as appropriate	1 lb 10 oz

BASIC METHOD FOR PREPARING PURÉE SOUPS

1 Soak the pulse vegetable in cold water for 12 hours approximately.
2 Place two-thirds of the stock in a saucepan or boiler.
3 Drain the water off the soaked pulse and add the pulse to the stock.
4 Bring to the boil, stirring as required to prevent burning.
5 Skim off the scum which forms on top of the liquid. Also remove any dried scum which forms on the sides of the pan.
6 Add the vegetables and herbs and slowly simmer. Skim and top up with additional stock as required during cooking.

continued overleaf

7   When cooked, pass the soup through a soup machine or liquidizer, then check consistency.

Note: When preparing green pea soup, remove the carrots prior to passing the soup through a soup machine or liquidizer.

Garnished purée soups

Purée soups may be garnished with a wide range of vegetables to produce individuality in taste, texture, appearance and nutritional value.

Suggested additions for purée soups

200 g/7 oz
{
Boiled rice, white or brown
Diced red and green peppers
Corn kernels
Diced root vegetables, e.g., carrots, turnips, etc.
Plain dried wholemeal bread croûtons
}

BROTHS
5 litres/8 ¾ pt, 25 portions

Mutton, chicken, pulse and farmers' broth

Each portion contains (garnish included):
114 kJ/27 kcal, 1.4 g fibre, <1 g fat.

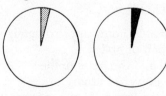

Energy
from fat 4%

Fibre
contribution

Basic recipe

5 litres	White stock*	8¾ pt
250 g	Diced onions	9 oz
250 g	Diced carrots	9 oz
250 g	Diced turnip	9 oz
120 g	Diced celery	4 oz
120 g	Diced leeks	4 oz
30 g	Chopped parsley	1 oz
10 g	Salt	2 tspns

*Use stock of appropriate flavour.

Mutton broth
Add 120 g/4 oz pearl barley.

Chicken broth
Add 120 g/4 oz brown rice.

Pulse broth
Add 150 g/5 oz mixed pulse vegetables: lentils, split peas and various types of beans.

Farmers' broth
Add 150 g/5 oz mixed pulse vegetables and cereals: haricot beans, lentils, brown rice and barley.

BASIC METHOD FOR PREPARING BROTHS

1 Soak the cereal or pulse in cold water for 12 hours.
2 Place the stock in a saucepan or boiler then bring to the boil.
3 Add any barley, marrow fat peas, lentils or beans and simmer for 1 hour approximately. Skim as required.
4 Add the onions, carrots, turnip and celery and simmer for 15 minutes approximately.
5 Add the leeks and any rice then simmer until cooked.
6 Check consistency and sprinkle with the chopped parsley when serving.

Suggested additions for broths

	Root fennel
	Corn kernels
200 g/7 oz	Beans
	Chick peas
	Brown rice
	Brown pasta

THICKENED VEGETABLE SOUPS AND CREAM SOUPS

5 litres/8¾ pt
25 portions

Artichoke, asparagus, cauliflower, celery, leek, mushrooms and onion

25 portions

Each portion contains:
260 kJ/61 kcal, 2 g fibre,
<1 g fat.

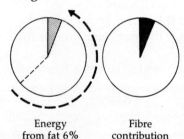

Energy   Fibre
from fat 6%   contribution

This cream will add 1 g of fat per portion and increase the energy from fat to 17%.

Basic recipe

This low-fat recipe produces soups with a surprisingly creamy flavour.

3¾ litres	White chicken stock	6 pt 12 fl. oz
350 g	Chopped onions	12 oz
120 g	Sliced celery	4 oz
120 g	Sliced white leeks	4 oz
	Sprig of thyme and bay leaf	
10 g	Salt	2 tspns

Thickening paste

120 g	Besan flour	4 oz
120 g	Plain flour	4 oz
1¼ litres	Cold milk (skimmed)	2 pt 4 oz

Cream soups

200 ml	Half cream	7 fl. oz

Artichoke soup
Add 1 kg/2 lb 3 oz artichokes to the basic soup recipe.

Asparagus soup
Add 2 kg/4 lb 6 oz asparagus trimmings to the basic recipe. The completed soup may be garnished with small pieces of cooked asparagus.

Cauliflower soup
Add 1 kg/2 lb 3 oz cauliflower pieces to the basic recipe. The completed soup may also be garnished with small florets of cooked cauliflower.

Celery soup
Increase the celery in the basic recipe to 1 kg/2 lb 3 oz. The completed soup may also be garnished with a dice of cooked celery.

Leek soup
Increase the leeks in the basic recipe to 1 kg/2 lb 3 oz.

Mushroom soup
Add 500 g/18 oz washed sliced mushrooms to the basic recipe. The soup may also be garnished with sliced cooked mushrooms.

Onion soup
Increase the onion in the basic recipe to 1 kg/2 lb 3 oz.

Soup of your choice
Add 1 kg/2 lb 3 oz vegetables of your choice to the basic recipe. Also garnish the completed soup with cooked vegetables of your choice.

BASIC METHOD FOR THICKENED VEGETABLE SOUPS

1 Wash and prepare the vegetables.
2 Place the stock in a saucepan or boiler, add the vegetables and bring to the boil.
3 Simmer for 20 minutes then whisk in the thickening paste.
4 Continue simmering until the vegetables are cooked, 20–30 minutes.
5 Purée or liquidize the soup.
6 Check consistency then add the garnish.
7 *Cream soups:* Blend through the cream just prior to service.

OTHER SOUPS

Mulligatawny

5 litres/8¾ pt
25 portions

Each portion contains:
394 kJ/93 kcal, 1.2 g fibre,
3 g fat.

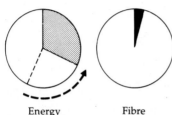

Energy   Fibre
from fat 32%   contribution

60 ml	Polyunsaturated oil	2 fl. oz
750 g	Chopped onions	1 lb 10 oz
30 g	Crushed garlic	1 oz
150 g	Curry powder	5 oz
120 g	Tomato purée	4 oz
250 g	Chopped cooking apples	9 oz
120 g	Mango chutney	4 oz
3 litres	Brown stock	5¼ pt
10 g	Salt	2 tspns

Thickening paste

120 g	Besan flour	4 oz
120 g	Plain flour	4 oz
1 litre	Cold brown stock	1¾ pt

Garnish

120 g	Cooked brown rice	4 oz

METHOD

1. Shallow fry the onions and garlic in the oil until golden brown.
2. Add the curry powder and continue cooking for 2–3 minutes.
3. Add the tomato purée, apples, chutney and brown stock and bring to the boil.
4. Allow to simmer for 15 minutes then add the thickening paste.
5. Continue simmering for a further 20 minutes until cooked.
6. Liquidize if desired then check consistency and add the garnish.

Tomato soup

5 litres/8¾ pt
25 portions

Each portion contains:
401 kJ/95 kcal, 2 g fibre,
3 g fat.

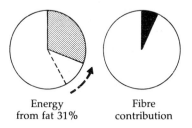

Energy
from fat 31%

Fibre
contribution

60 ml	Polyunsaturated oil	2 fl. oz
500 g	Diced onions	18 oz
500 g	Diced carrots	18 oz
250 g	Diced celery	9 oz
250 g	Chopped leeks	9 oz
10 g	Crushed garlic	½ oz
	Sprig of thyme, 2 bay leaves and parsley stalks	
200 g	Lean bacon scraps	7 oz
500 g	Tomato purée	18 oz
4 litres	Stock	7 pt
10 g	Salt (if not bacon stock)	2 tspns

Thickening paste

130 g	Besan flour	5 oz
130 g	Plain flour	5 oz
1 litre	Cold Stock	1¾ pt

METHOD

1 Sweat the vegetables and bacon in the oil until the vegetables are soft, 8–12 minutes approximately.
2 Add the tomato purée and stock and bring to the boil.
3 Allow to simmer for 15 minutes then whisk in the thickening paste.
4 Continue cooking for a further 20 minutes until the soup is cooked.
5 Withdraw the bay leaves and parsley stalks (or bouquet garni containing the herbs) and check consistency.
6 Purée or liquidize the soup if desired.

Fresh tomato soup with basil and crisp vegetables

5 litres/8¾ pt
25 portions

Each portion contains:
385 kJ/91 kcal, 4 g fibre,
3 g fat.

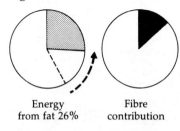

| Energy from fat 26% | Fibre contribution |

60 ml	Polyunsaturated oil	2 fl. oz
250 g	Chopped onions	9 oz
250 g	Diced carrots	9 oz
120 g	Sliced leeks	4 oz
120 g	Diced celery	4 oz
10 g	Garlic	½ oz
30 g	Fresh basil	1 oz
	Sprig of thyme, 2 bay leaves and parsley stalks	
150 g	Tomato purée	5 oz
4 kg	Ripe tomatoes (sliced)	8 lb 12 oz
1½ litres	Chicken stock	2 pt 12 fl. oz
10 g	Salt	2 tspns

Thickening paste

60 g	**Besan flour**	2 oz
60 g	**Plain flour**	2 oz
250 ml	**Cold stock**	9 fl. oz

Garnish

200 g	**Diced green peppers**	7 oz
200 g	**Sweetcorn kernels**	7 oz
200 g	**Cooked brown rice**	7 oz
10 g	**Fresh chopped basil**	½ oz

METHOD

1 Sweat the vegetables and herbs in the oil until soft, 8–10 minutes.
2 Add the tomato purée, ripe tomatoes (including seeds) and chicken stock, bring to the boil.
3 Simmer for 15 minutes, add the thickening paste.
4 Cook for a further 20 minutes until all the vegetables are cooked, then pass the soup through a liquidizer.
5 Bring the soup back to the boil, add the garnish and simmer for 5 minutes approximately.
6 Check consistency and serve.

Vichysoisse with peppers and corn kernels

5 litres/8¾ pt
25 portions

Each portion contains:
344 kJ/81 kcal, 3 g fibre,
<1 g fat.

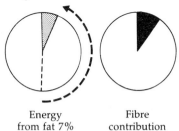

| Energy | Fibre |
| from fat 7% | contribution |

300 g	Diced onions	10½ oz
1 kg	Chopped white leeks	2 lb 3 oz
1½ kg	Peeled and sliced potatoes	3 lb 4 oz
	Sprig of thyme, 2 bay leaves and parsley stalks	
3 litres	White chicken stock	5¼ pt
120 ml	Half cream (optional)	4 fl. oz
10 g	Salt	2 tspns

Garnish

30 g	Chopped chives	1 oz
30 g	Chopped parsley	1 oz
200 g	Sweetcorn kernels	7 oz
200 g	Diced red peppers	7 oz

METHOD

1 Place the stock in a saucepan or boiler, then add the onions, leeks, potatoes and herbs.
2 Bring to the boil and simmer until the vegetables are cooked, 30–40 minutes. Remove the bay leaves and parsley stalks.
3 Pass through a soup machine or liquidize, allow to cool.
4 Thoroughly chill the soup, then check consistency.
5 Stir through the cream (if used), add the vegetable garnish.

Note: The cream is included in the calculations.

*Gazpacho★★

5 litres/8¾ pt
25 portions

Each portion contains:
422 kJ/101 kcal, 2 g fibre,
8 g fat.

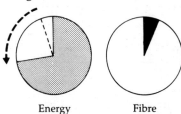

| Energy | Fibre |
| from fat 73% | contribution |

*Symbols denote a recipe with a percentage energy higher than 35%; see Chapter 3, page 62.

300 g	Chopped onions	10½ oz
60 g	Crushed garlic	2 oz
500 g	Chopped green peppers	18 oz
1½ kg	Sliced tomatoes (with seeds)	3 lb 4 oz
750 g	Cucumber	1 lb 10 oz
To taste	Ground black pepper	To taste
30 g	Ground cumin	1 oz
60 ml	Lemon juice	2 fl. oz
120 ml	Vinegar	4 fl. oz
200 ml	Polyunsaturated oil	7 fl. oz
10 g	Salt	2 tspns
	Ice water (to adjust consistency)	

Garnish

250 g	Finely chopped onion	9 oz
250 g	Diced cucumber	9 oz
120 g	Diced tomato flesh	4 oz
250 g	Diced red and green peppers	9 oz

METHOD

1 Liquidize the vegetables with the lemon juice, vinegar and oil.
2 Add the spices and correct the consistency with the iced water.
3 Chill thoroughly, add the garnish just prior to service.

Fenugreek soup ★★

5 litres/8 ¾ pt
25 portions
Each portion contains:
337 kJ/80 kcal, 1 g fibre,
4 g fat.

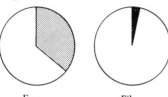

Energy
from fat 36%

Fibre
contribution

30 ml	Polyunsaturated oil	1 fl. oz
200 g	Chopped onions	7 oz
5 g	Fenugreek	3 tspns
1.2 litre	Low-fat natural yogurt	2 pt 2 fl. oz
3 litres	Chicken stock	5 pt 4 oz
80 g	Chopped walnuts	3 oz
30 g	Chopped fresh parsley	1 oz
To taste	Ground black pepper	To taste
10 g	Salt	2 tspns

Thickening paste

130 g	Besan flour	5 oz
130 g	Plain flour	5 oz
800 ml	Cold chicken stock	1 pt 8 fl. oz

METHOD

1   Sweat the onion in the oil for 4–5 minutes, add the fenugreek. Sweat for a further 3 minutes approximately.
2   Add the chicken stock and bring to the boil, simmer for 5 minutes approximately.
3   Prepare the thickening paste, add to the soup and reboil.
4   Simmer for 8–10 minutes, then mix in the yogurt.
5   Add the walnuts, parsley and pepper. Check temperature and consistency.

Clam and bean chowder★★

5 litres/8¾ pt
25 portions

Each portion contains:
296 kJ/71 kcal, 2 g fibre,
3 g fat.

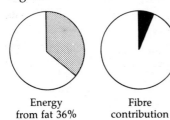

Energy
from fat 36%

Fibre
contribution

60 ml	Polyunsaturated oil	2 fl. oz
300 g	Diced onions	10½ oz
300 g	Diced white leeks	10½ oz
300 g	Diced potatoes	10½ oz
300 g	Cooked small haricot beans	10½ oz
250 g	Diced cooked clams	9 oz
4½ litres	Fish stock	7 pt 18 fl. oz
80 ml	Half cream	3 fl. oz
120 g	Wholemeal cracker biscuits	4 oz
10 g	Chopped fresh herbs	½ oz
10 g	Salt	2 tspns

METHOD

1 Sweat the onions and leeks in the oil until soft, 8–10 minutes.
2 Add the stock and bring to the boil.
3 Add the potatoes and simmer until the potatoes are cooked, 10 minutes approximately.
4 Add the cooked beans and clams and reheat thoroughly.
5 Check consistency, then blend in the cream.
6 Break the biscuits into small pieces, stir through the soup when serving.

PASTA DISHES

Pastas, like soups, may be used as part of a meal or as a complete meal by themselves. Pasta dishes may be made in many forms and offer a good source of fibre, especially when wholemeal pasta is used.

Cookery hints

- Always cook in plenty of boiling water to avoid starchy pasta which sticks together. If pasta is sticky after cooking, refresh in hot water.
- Do not toss pastas in oil or butter after cooking; simply mix the pasta with any garnish and sauce in a careful manner.
- When preparing baked or gratinated pasta, for example lasagne, use only a thin coating of polyunsaturated oil on surfaces of casseroles or cooking/service containers.

Pasta with bolonaise and vegetable sauce

Examples: spaghetti, macaroni or noodles

10 portions

Each portion contains:
1659 kJ/395 kcal, 6 g fibre,
9 g fat.

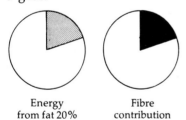

| Energy from fat 20% | Fibre contribution |

600 g	Dried pasta	1 lb 5 oz
1½ litres	Bolonaise and vegetable sauce (page 80)	2 pt 12 fl. oz
To taste	Black pepper	To taste
120 g	Grated low-fat cheese	4 oz

METHOD

1 Cook the pasta in plenty of boiling water then drain thoroughly (refresh and reheat if required).
2 Mix together the hot sauce, pepper and pasta and sprinkle the grated cheese over the top.
3 Alternatively, serve the pasta, cheese and sauce separately.

Pasta with vegetables and pine nuts

Examples: spaghetti, macaroni or noodles

10 portions

Each portion contains:
1401 kJ/333 kcal, 6 g fibre,
8 g fat.

600 g	Dried pasta	1 lb 5 oz
1.2 litres	Tomato sauce (page 81)	2 pt 2 fl. oz
300 g	Diced onions	10½ oz
300 g	Diced and blanched carrots	10½ oz
150 g	Diced and blanched celery	5 oz
150 g	Sliced mushrooms	5 oz
90 g	Pine nuts	3 oz
To taste	Chopped thyme	To taste

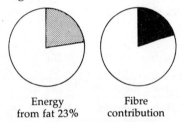

Energy
from fat 23%

Fibre
contribution

METHOD

1 Cook the pasta ready for service.
2 Add the vegetables to the sauce and simmer for 2–3 minutes. Check the consistency and adjust with a little stock if required.
3 Blend the pasta and sauce together, then sprinkle with the pine nuts and chopped thyme. Alternatively, serve the sauce and pasta separately.

Vegetable dumplings in Mexican chilli sauce

10 portions

Each portion contains:
1350 kJ/342 kcal, 6 g fibre,
12 g fat.

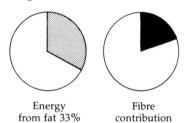

Energy
from fat 33%

Fibre
contribution

Dumplings

200 g	Plain flour	7 oz
200 g	Wholemeal flour	7 oz
120 g	Polyunsaturated margarine	4 oz
200 g	Cooked brown rice	7 oz
120 g	Finely chopped onions	4 oz
60 g	Chopped celery	2 oz
60 g	Chopped peppers	2 oz
60 g	Diced carrots	2 oz
60 g	Chopped sweet corn kennels	2 oz
60 g	Chopped blanched green beans	2 oz
To taste	Black pepper	To taste
60 ml	Cold water	2 fl. oz

Mexican chilli sauce

30 ml	Polyunsaturated oil	1 fl. oz
120 g	Chopped onions	4 oz
10 g	Crushed garlic	½ oz
60 g	Diced red peppers	2 oz
60 g	Diced green peppers	2 oz
10 g	Ground cumin	½ oz
5 g	Ground coriander	2 tspns
To taste	Ground black pepper	To taste
To taste	Chilli pepper	To taste
To taste	Cayenne pepper	To taste
500 g	Tinned tomatoes	1 lb 2 oz
450 ml	Tomato sauce (page 81)	16 fl. oz

METHOD

A Preparing the dumplings
1 Mix together the flours and black pepper.
2 Lightly rub in the margarine.
3 Mix through all the vegetables and cooked brown rice.

continued overleaf

4   Add and mix in the cold water until a stiff dough is formed.
5   Shape the mixture into small balls.
6   Add the balls to simmering water and poach until cooked, 6–8 minutes.
7   Remove and drain.

B   *Preparing the sauce*
8   Sweat the vegetables in the oil until soft, 3–4 minutes.
9   Add the spices and continue cooking for 2–3 minutes.
10 Add the tinned tomatoes and low-fat tomato sauce and bring to the boil.
11 Simmer until a thin coating consistency is achieved.

C   *Service*
12 Carefully mix together the hot dumplings and sauce.

Note: The vegetables in the dumplings must be finely chopped to prevent them from splitting away from the dough mixture.

Vegetable lasagne

10 portions

Each portion contains:
1365 kJ/324 kcal, 9 g fibre,
11 g fat.

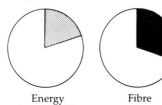

| Energy | Fibre |
| from fat 20% | contribution |

660 g	*Wholemeal lasagne (pasta recipe on page 196)	1 lb 7 oz
200 g	Chopped onions	7 oz
120 g	Diced blanched green beans	4 oz
120 g	Diced blanched celery	4 oz
120 g	Diced blanched turnips	4 oz
300 g	Diced tomatoes	10½ oz
120 g	Sweetcorn kernels	4 oz
120 g	Sliced mushrooms	4 oz
200 g	Blanched and drained spinach	7 oz
800 ml	Low-fat Béchamel (page 73)	1 pt 8 fl. oz
150 g	Low-fat cheddar-type cheese	5 oz
To taste	Ground black pepper	To taste

*Quantity stated is for fresh pasta. Use half the stated quantity when using dried pasta.

METHOD

1 Boil the lasagne in plenty of boiling water, refresh and drain.
2 Lightly oil an earthenware dish, place a layer of cooked lasagne over the bottom.
3 Coat with a layer of Béchamel, then add a layer of vegetables. Use half the quantity of vegetables (except the spinach).
4 Place a second layer of lasagne on top and coat with Béchamel.
5 Place on the spinach and cover with a further layer of lasagne.
6 Repeat step 3, adding Béchamel and the remaining vegetables.
7 Cover with a final layer of lasagne and coat with Béchamel.
8 Grate the cheese and sprinkle over the lasagne. Place in a moderate oven to thoroughly reheat and gratinate the surface.

Lasagne bolonaise with vegetables

Prepare as vegetable lasagne with the following amendments:

1 Replace the wholemeal lasagne with plain lasagne (page 196) and omit the spinach.
2 Use 400 ml Béchamel sauce and 500 ml bolonaise sauce (page 79). Use alternate coatings of each sauce when layering the pasta.

Pasta and potato bake

Examples: macaroni, tagliatelle and lasagne.
10 portions

Each portion contains:
1286 kJ/304 kcal, 5 g fibre, 7 g fat.

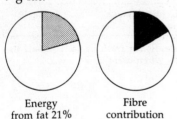

Energy from fat 21%   Fibre contribution

200 g	Pasta (dried)	7 oz
900 g	Blanched sliced potatoes (with skins, 5 mm/¼ inch slices)	2 lb
30 ml	Polyunsaturated oil	1 fl. oz
450 g	Sliced onions	1 lb
10 g	Crushed garlic	½ oz
120 g	Sliced mushrooms	4 oz
1 litre	Low-fat cheese sauce (page 74)	1¾ pt
60 g	Wholemeal breadcrumbs	2 oz
30 g	Grated Parmesan	1 oz

METHOD

1 Cook and drain the pasta.
2 Sweat the onions and garlic in the oil under tender.
3 Add the mushrooms and sweat for a further 2–3 minutes.
4 Carefully bind the pasta, potatoes and vegetables with the sauce and place into an ovenproof service dish.
5 Clean round the sides of the dish if required, then sprinkle the breadcrumbs and cheese over the top.
6 Place in a moderate oven to thoroughly heat through and develop a good colour on the surface of the bake.

Macaroni and lentil bake

10 portions

Each portion contains:
912 kJ/217 kcal, 5 g fibre,
6 g fat.

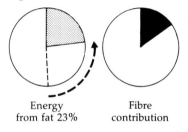

Energy
from fat 23%

Fibre
contribution

200 g	Macaroni	7 oz
200 g	Lentils (brown or green)	7 oz
30 ml	Polyunsaturated oil	1 fl. oz
250 g	Diced onions	9 oz
10 g	Crushed garlic	½ oz
120 g	Diced green peppers	4 oz
250 g	Sliced courgettes (with skins)	9 oz
250 g	Diced peeled aubergines	9 oz
600 ml	Tomato sauce	1 pt
5 g	Ground coriander	2 tspns
To taste	Ground black pepper	To taste
30 g	Low-fat cheddar-type cheese (grated)	1 oz
30 g	Wholemeal breadcrumbs	1 oz

METHOD

1 Cook and drain the macaroni.
2 Cook and drain the lentils (do not overcook).
3 Sweat the onions and garlic in the oil for 2–3 minutes.
4 Add the coriander, black pepper, green peppers, courgettes, aubergines and tomato sauce.
5 Simmer for 3–4 minutes (stirring occasionally to avoid burning).
6 Add the macaroni and lentils and carefully mix all the ingredients together.
7 Place the mixture into an ovenproof serving dish, then sprinkle over the surface with the cheese and breadcrumbs.
8 Heat through and gratinate in a moderate oven.

MEAT, POULTRY AND GAME STEWS

All meat, game and poultry stews and braisings are easily extended with vegetables. This also applies to ground meat or minced dishes such as chilli con carne and shepherd's pie.

Cookery hints

- Stews and braisings are often garnished with shallow-fried or buttered items, for example glazed button onions, glazed or buttered carrots and turnips, and fried cubes of bacon (lardons). Plainly cooked items (steamed or boiled) provide the ideal marriage with all stews and braised dishes.

BROWN STEWS

10 portions

Each portion contains:
948 kJ/225 kcal, 5 g fibre,
8 g fat.

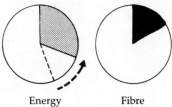

| Energy | Fibre |
| from fat 31% | contribution |

Basic recipe

30 ml	Polyunsaturated oil	1 fl. oz
150 g	Chopped onion	5 oz
150 g	Diced carrots	5 oz
80 g	Diced celery	3 oz
80 g	Diced leeks	3 oz
80 g	Tomato purée	3 oz
1 litre	Brown stock	1¾ pt
5 g	Salt	½ tspn
60 g	Wholemeal flour	2 oz
250 ml	Cold brown stock	9 fl. oz

Types of stew

Brown beef stew:	1 kg	Lean stewing beef	2 lb 3 oz
Brown veal stew:	1 kg	Lean stewing veal	2 lb 3 oz
Brown mutton stew:	1 kg	Lean stewing mutton★★	2 lb 3 oz
Brown rabbit stew:	1 kg	Rabbit flesh	2 lb 3 oz

METHOD

1 Cut the meat into 25 mm/1 inch cubes.
2 Heat the oil in a suitable saucepan.
3 Add the meat and sear the flesh until brown.

4 Add the basic vegetables and sweat for 8–10 minutes.
5 Add any spice stated in the recipe and sweat for 2–3 minutes.
6 Stir in the wholemeal flour to coat the meat and vegetables.
7 Mix in the tomato purée.
8 Mix in the brown stock and slowly simmer until almost cooked.
9 Throughout the cooking period, skim as required and top up with additional stock when necessary.
10 Check consistency and serve.

Suggested additions for the above stews

300 g/10 oz	Corn kernels	Butter beans
	Diced celeriac	French beans
	Diced celery	Peas
	Diced peppers	Bamboo shoots
	Lima beans	Palm hearts
	Red beans	Salsify

DERIVATIVE DISHES OF BROWN STEW

Rich brown stew in red wine sauce

Same as the basic recipe but replace 120 ml/4 fl. oz of the brown stock with red wine.

Beef in ale

Same as the basic recipe but with the following amendments:

1 Replace 300 g/10½ oz beef with lean bacon cut into strips.
2 Replace 300 ml/10½ fl. oz of the brown stock with beer.

Note: Bacon, which is salty, should be placed in cold water, brought to the boil and cooked for a short period to remove excess salt.

Beef and vegetable goulash

Same as the basic recipe but add 60 g/2 oz paprika to the stew at step 5. Also add a garnish of 200 g/7 oz of diced potatoes to the stew just prior to completion of cooking and the same quantity of cooked wholemeal macaroni to the stew when cooked.

Green pepper and red bean beef goulash

Each portion contains:
1023 kJ/243 kcal, 5 g fibre,
9 g fat.

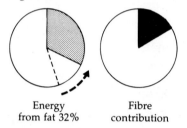

Energy
from fat 32%

Fibre
contribution

Same as the basic recipe but add 60 g/2 oz paprika to the stew at step 5. Also add a garnish of 200 g/7 oz diced green peppers to the stew just prior to completion of cooking and the same quantity of cooked red beans when cooked.

Beef ragout with courgettes and butter beans

Same as the basic recipe, but add 200 g/7 oz sliced courgettes to the stew just prior to completion of cooking and add the same quantity of cooked butter beans when cooked.

Navarin of lamb with peas and braised brown rice

Same as the basic recipe, but add 200 g/7 oz peas to the stew just prior to completion of cooking. Serve with braised brown rice (page 128.)

Beef olives

10 portions

Each portion contains:
1203 kJ/286 kcal, 4 g fibre,
9 g fat.

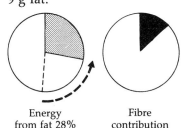

Energy
from fat 28%

Fibre
contribution

1 kg	Topside of beef (lean)	2 lb 3 oz
30 ml	Polyunsaturated oil	1 fl. oz
150 g	Chopped onions	5 oz
150 g	Diced carrots	5 oz
150 g	Diced celery	5 oz
60 g	Diced leeks	2 oz
700 ml	Low-fat brown sauce (page 76)	1 1/4 pt
5 g	Salt	1 tspn

Stuffing

200 g	Chopped onions	7 oz
200 g	Diced green peppers	7 oz
120 g	Wholemeal breadcrumbs	4 oz
120 g	Tomato purée	4 oz
2 g	Mixed herbs	2 tspns

METHOD

1 Prepare the stuffing by mixing together all the ingredients.
2 Slice the topside into 10 even slices (or 20 medium slices).
3 Prepare the olives: lay the stuffing along the centre of each slice then neatly roll up (securing with string is not required).
4 Shallow fry the vegetables in a suitable cooking utensil until lightly coloured.
5 Lay the olives on the bed of vegetables, cover with the sauce.
6 Cover with a lid then cook in a moderate oven until tender (do not stir the olives during cooking). Skim sauce as and when required.
7 Place the olives on the serving dish (or storage container). Check the consistency of the sauce.
8 Pour the sauce over the olives and serve.

Turkey olives

Prepare the same as beef olives, replacing the beef slices with slices of turkey breast.

Chilli con carne with crisp vegetables

10 portions

Each portion contains:
988 kJ/234 kcal, 7 g fibre,
8 g fat.

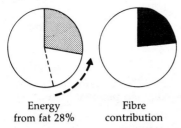

Energy from fat 28%   Fibre contribution

30 ml	Polyunsaturated oil	1 fl. oz
140 g	Chopped onions	5 oz
5 g	Crushed garlic	2 cloves
750 g	Lean minced beef	1 lb 10 oz
*10 g	Chilli powder	½ oz
15 g	Paprika	½ oz
5 g	Ground cumin	2 tspns
60 g	Plain flour	2 oz
30 g	Tomato purée	1 oz
300 g	Tinned tomatoes (with juice)	10½ oz
600 ml	Bean cooking liquor (from beans)	1 pt
250 g	Dried red beans	9 oz
5 g	Salt	1 tspn

Garnish

400 g	Diced onions Diced green peppers Diced red peppers Sweetcorn Diced celery	14 oz

*Take care when using hot chilli powder.

METHOD

1 Soak the beans for 6–12 hours in cold water then cook in brown stock.
2 Heat the oil in a saucepan, shallow fry the onions and garlic until soft, 3–4 minutes.
3 Add the mince and shallow fry until lightly coloured.
4 Add the spices and cook slowly for 2–3 minutes.
5 Mix in the plain flour, then allow to cook for 3–4 minutes.
6 Add the tomato purée and tinned tomatoes.
7 Blend in the cooking liquor from the beans and bring to the boil.
8 Slowly simmer until cooked.
9 Add the beans and vegetables and cook for a further 4–5 minutes. Avoid over-cooking and leave the vegetables crisp and tender.

Chinese pork with pineapple, peppers and water chestnuts

10 portions

Each portion contains:
1412 kJ/336 kcal, 1.5 g fibre, 11 g fat.

30 ml	Polyunsaturated oil	1 fl. oz
1 kg	Diced lean stewing pork	2 lb 3 oz
150 g	Sliced onions	5 oz
250 g	Sliced green and red peppers	9 oz
170 g	Fresh pineapple	6 oz
60 ml	Fresh pineapple juice	2 fl. oz
150 g	Water chestnuts	5 oz
30 ml	Light soy sauce	1 fl. oz
30 ml	Sherry	1 fl. oz
300 ml	Chicken stock (to cover)	10½ fl. oz
20 g	Cornflour (to thicken)	½ oz
5 g	Salt	1 tspn

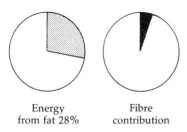

Energy from fat 28%   Fibre contribution

Accompaniment

*1 kg	Boiled rice	2 lb 3 oz

*The fibre content will be increased to 5 g per portion if brown rice is used as an accompaniment.

METHOD

1 Peel the pineapple, cut the flesh into dice and reserve the juice (squeeze the skins and cut fruit).
2 Marinade the pork in the pineapple juice, sherry and soy sauce for 30 minutes.
3 Remove the pork from the marinade and thoroughly drain.
4 Heat the oil in a suitable cooking utensil and sear the pork.
5 Add the marinade and chicken stock to barely cover the flesh.
6 Simmer until the meat is tender.
7 Add the peppers, onions, pineapple and water chestnuts to the stew, thicken with the cornflour diluted in a little cold water.
8 Complete the cooking, keeping the vegetables crisp in texture, 3 minutes approximately.
9 Serve with the boiled rice.

Blanquette of lamb

10 portions

Each portion contains:
1663 kJ/394 kcal, 4 g fibre,
11 g fat.

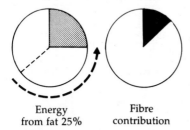

Energy   Fibre
from fat 25%   contribution

1 kg	Diced lean stewing lamb	2 lb 3 oz
2	Bay leaves	2
250 g	Diced onions	9 oz
250 g	Button mushrooms	9 oz
120 g	Cooked butter beans	4 oz
1 litre	White stock (to cover)	1¾ pt
200 g	Fromage frais	7 oz
30 g	Besan flour	1 oz
30 g	Plain flour	1 oz
60 ml	Lemon juice	2 fl. oz
5 g	Salt	1 tspn

Accompaniment

1 kg	Cooked brown rice	2 lb 3 oz

METHOD

1 Cover the lamb with cold water, bring to the boil, then refresh in cold running water. Drain ready for cooking.
2 Cover with the stock, add the bay leaves, bring to the boil and simmer until almost cooked. Top up with stock and skim as required.
3 Mix together the fromage frais and the two flours to a paste (fromage manié).
4 Remove the bay leaves and whisk in the thickening paste, then reboil to thicken.
5 Add the onions and cook for 4–5 minutes.
6 Add the mushrooms and simmer until cooked.
7 Add the butter beans and heat through.
8 Finish with the lemon juice and check consistency.

Irish stew with haricot beans

10 portions

Each portion contains:
1150 kJ/272 kcal, 5 g fibre,
9 g fat.

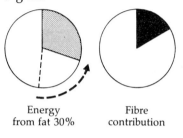

Energy
from fat 30%

Fibre
contribution

1 kg	Diced lean stewing lamb	2 lb 3 oz
800 kg	Peeled sliced potatoes	1 lb 12 oz
200 g	Peeled diced potatoes	7 oz
150 g	Diced celery	5 oz
150 g	Diced leeks	5 oz
150 g	Shredded white cabbage	5 oz
150 g	Diced turnips	5 oz
120 g	Cooked haricot beans	4 oz
1 litre	White stock (to cover)	1¾ pt
5 g	Salt	1 tspn

METHOD

1 Cover the lamb with cold water, bring to the boil, then refresh in cold running water. Drain ready for cooking.
2 Cover with the stock and bring to the boil.
3 Add the sliced potatoes and celery and simmer for 45 minutes approximately (until the potatoes break down and form a sauce).
4 Add the diced potatoes, turnips, cabbage and leeks and complete the cooking. Top up with additional stock and skim off surface face as and when required.
5 Add the haricot beans and allow to heat through.
6 Check consistency and serve.

Red bean and tomato moussaka

10 portions

Each portion contains:
989 kJ/235 kcal, 9 g fibre,
7 g fat.

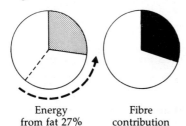

Energy   Fibre
from fat 27%   contribution

1 kg	Sliced aubergines	2 lb 3 oz
500 g	Sliced tomatoes	1 lb 2 oz
250 g	Cooked red beans	9 oz
1½ litres	Bolonaise and vegetable sauce (page 80)	2 pt 12 fl. oz
30 g	Breadcrumbs (wholemeal)	1 oz
60 g	Low-fat cheddar-type cheese	2 oz
To taste	Ground black pepper	To taste
	Chopped parsley	

METHOD

1 Blanch the aubergines in boiling water for 30 seconds.
2 Lightly oil the base of a suitable cooking utensil, cover with a layer of blanched aubergines (half the quantity of aubergines approximately).
3 Place half the quantity of red beans over the aubergines, cover with the bolonaise and vegetable sauce.
4 Place the remaining beans on top, neatly cover with the remaining aubergines and sliced tomatoes.
5 Sprinkle the breadcrumbs and cheese over the surface.
6 Gratinate in a hot oven until golden brown and thoroughly reheated. Sprinkle with chopped parsley.

Beef and vegetable pie

10 portions

Each portion contains:
1060 kJ/251 kcal, 4 g fibre,
7 g fat.

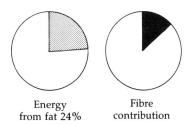

| Energy from fat 24% | Fibre contribution |

30 ml	Polyunsaturated oil	1 fl. oz
800 g	Diced lean beef	1 lb 12 oz
60 g	Tomato purée	2 oz
900 ml	Brown beef stock	1 pt 12 fl. oz
150 g	Diced onions	5 oz
150 g	Diced blanched carrots	5 oz
120 g	Diced blanched swedes	4 oz
120 g	Sliced blanched celery	4 oz
150 g	Blanched cauliflower pieces	5 oz
120 g	Sliced courgettes	4 oz
150 g	Sweetcorn kernels	5 oz
120 g	Sliced mushrooms	4 oz
5 g	Salt	1 tspn

Thickening paste

60 g	Plain flour	2 oz
60 ml	Cold stock or water	2 fl. oz

Topping

900 g	Mashed potatoes	2 lb

METHOD

1 Heat the oil in a saucepan then sear the flesh until brown. Drain off any fat which is present.
2 Add the tomato purée and stock and bring to the boil.
3 Simmer until almost cooked. Top up with stock and skim off surface fat as necessary.
4 Add the thickening paste and reboil.
5 Add all the vegetables and cook for a short period, keeping the vegetables crisp in texture.
6 Place the mixture in a suitable cooking/serving dish, then cover with the potatoes. Brush over the surface of the potato with a little milk.
7 Cook in a moderately hot oven until golden brown and reheated.

Cottage and farm pie

10 portions

Each portion contains:
880 kJ/207 kcal, 5 g fibre,
5 g fat.

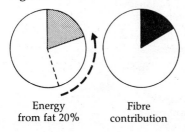

Energy   Fibre
from fat 20%   contribution

30 ml	Polyunsaturated oil	1 fl. oz
600 g	Lean minced beef	1 lb 5 oz
30 g	Wholemeal flour	1 oz
60 g	Tomato purée	2 oz
450 ml	Brown stock	16 fl. oz
150 g	Diced onions	5 oz
150 g	Diced blanched carrots	5 oz
150 g	Sliced blanched celery	5 oz
70 g	Diced blanched swede	2 oz
70 g	Diced blanched turnip	2 oz
70 g	Diced courgettes	2 oz
150 g	Peeled diced marrow	5 oz
150 g	Cooked red beans	5 oz
150 g	Diced blanched green beans	5 oz
5 g	Salt	1 tspn

Topping

900 g	Mashed potatoes	2 lb

METHOD

1 Heat the oil in a saucepan then fry off the mince until brown.
2 Add the flour and mix through the mince.
3 Cook over a low heat for 3–4 minutes, stirring occasionally.
4 Add the tomato purée then stir in the stock.
5 Bring to the boil, simmer for 30 minutes approximately. Top up with stock and skim off surface fat as necessary.
6 Add the vegetables and cook for a short period (8 minutes approximately), keeping the vegetables crisp in texture.
7 Place the mixture in a suitable cooking/serving dish then cover with the potatoes.
8 Lightly brush over the surface of the potatoes with a little milk. Cook in a moderate oven until golden brown and thoroughly reheated.

VEGETABLES

COOKING VEGETABLES

Vegetables contain valuable nutrients which are often unnecessarily destroyed by bad cookery practices. Also, many cookery books make suggestions on how to cook vegetables which only *increase the loss* of nutrients, these suggestions being based on popular cookery myths which have no foundation in fact (see cookery hints on cooking root vegetables below). The aim of any good cook when cooking vegetables should be to achieve the best balance between flavour, colour, texture and nutritional value.

Cooking green leaf and leguminous vegetables

This comprises all *fresh and frozen* peas, sugar peas, French beans, runner beans, lima beans, cabbage, brussel sprouts and spinach, etc.

Cookery hints

- When preparing the vegetables, never soak the vegetables in cold water as valuable water soluble vitamins will leach out and be lost.
- Always cut the vegetables with a sharp knife. A blunt knife bruises the vegetables and causes a loss of vitamin C brought about by oxidative enzymes.
- Always start the cooking in a minimum quantity of boiling water to reduce nutrient and flavour loss; but do not use too little water as it may boil dry and scorch the vegetables. *Never start the cooking in cold water.*
- Cover with a lid to allow steam to be used in the cooking process. Rapid cooking with a lid on the cooking utensil results in the shortest possible cooking time—*prolonged cooking is the most significant factor with regard to loss of texture, colour, flavour and nutritive value.* Where possible cook the vegetables in a high-speed steamer; this is a desirable alternative to boiling the vegetables.
- Toss and inspect the vegetables during cooking—this promotes even cooking and allows volatile acids, which cause deterioration in the colour of green vegetables, to escape.
- Cook the vegetables in small batches to keep hot storage to a minimum.
- Test for cooking regularly to avoid over-cooking. Never rely on times stated in cookery books as they are only rough guides.
- Remember that frozen vegetables require very little cooking and are easily over-cooked.
- Do not follow the traditional procedure of tossing vegetables in butter after cooking.

Cooking root vegetables

This comprises carrots, turnips, swedes, parsnips and potatoes, etc.

Cookery hints

Most of the points above apply to root vegetables. However, the vegetables should be covered with water (but still a minimum quantity of water should be used) to ensure even cooking.

- Disregard the traditional practice of starting vegetables grown under the ground in cold water. This practice reduces nutritional value and is a cookery myth which should be ignored. Always start the cooking in boiling or very hot water where possible. However, be careful when adding the vegetables to boiling water; burns and scalds may easily result, especially when large quantities of root vegetables are involved.

Cooking flower and stem vegetables

This comprises asparagus, cauliflower, broccoli and celery, etc.

Follow all the points for green leaf vegetables, but like root vegetables above use only enough boiling water to cover and cook the vegetables.

Cooking pulse vegetables

This comprises all dried beans, peas and lentils—see Chapter 2, pages 56–57 for varieties, cooking times and approximate weight increases after soaking and cooking.

Cookery hints

- Wash and soak in cold water for 8 hours prior to cooking. Alternatively, bring to the boil and simmer for 2–3 minutes, then leave to soak for 1 hour in hot water.
- Rinse thoroughly after soaking in cold water.
- Do not add salt or acid (e.g., lemon juice) to the cooking water as it toughens the skins.
- Pressure cooking at 105 kN/m^2 (15 lb psi) (pounds per square inch) only requires one-quarter to one-third of the boiling time (see page 56).
- IMPORTANT: Dried beans contain a toxin (haemagglutinin) which can cause gastroenteritis. *However, the danger is eliminated if the beans are rapidly boiled for 10 minutes then thoroughly cooked through.*
- Cooked pulses freeze well and may be defrosted and used as required.

VEGETABLE STEWS

Combinations of vegetables can be used not just as accompaniments to meat, but as complete meals on their own. They produce colourful filling dishes with an inbuilt fibre content; the following suggestions have been adapted from continental cuisine.

Ratatouilles

These are vegetable stews based on courgettes, or marrow when in season, aubergines, green peppers and tomatoes. Ratatouilles can be used hot as a vegetable, cold as a salad or served as a main course with the addition of diced meat or chicken. Traditional recipes suggest frying the vegetables in olive oil but the aubergine in particular will soak up much of the oil. However, by cooking the tomatoes and onions first you can make a sauce in which to cook the remaining ingredients and achieve a dish much lower in fat.

Basic recipe for ratatouilles

10 portions

Each portion contains: 272 kJ/64 kcal, 4 g fibre, 2 g fat.

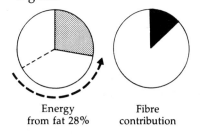

| Energy from fat 28% | Fibre contribution |

15 ml	Olive oil	½ fl. oz
500 g	Sliced onions	1 lb 2 oz
750 g	Chopped fresh or tinned tomatoes	1 lb 10 oz
500 g	Peeled sliced aubergines	1 lb 2 oz
500 g	Sliced green peppers	1 lb 2 oz
500 g	Sliced courgettes	1 lb 2 oz
10 g	Crushed garlic (3 large cloves)	½ oz
5 g	Ground coriander	2 tspns
To taste	Ground black pepper	To taste
	Fresh chopped parsley	
5 g	Salt	1 tspn

BASIC METHOD

1 Heat the oil in a saucepan, add the onions and sweat with the lid on until soft (4–5 minutes). Add the coriander and cook for 2–3 minutes.

continued overleaf

2   Add the tomatoes and cook until they soften and form a sauce.
3   Add the aubergines and peppers and cook with lid on for 15 minutes approximately.
4   Add the courgettes and continue cooking until all the vegetables are tender (but still crisp in texture).
5   Add the parsley and check consistency. Serve hot or cold.

Sicilian ratatouille★★

10 portions

Each portion contains:
586 kJ/139 kcal, 5 g fibre, 6 g fat.

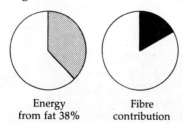

Energy   Fibre
from fat 38%   contribution

Replace the courgettes with 500 g/1 lb 2 oz sliced blanched celery (1 medium head of celery). Add the celery at the same time as the aubergines. Towards the end of the cooking time, add 200 g/7 oz diced rollmop herrings (5 fillets approximately) and 60 g/2 oz capers. Chill and serve as a starter with wholemeal bread.

Note: Anchovy fillets may be used in place of rollmop herring (use 8 anchovy fillets).

Mushroom ratatouille

10 portions

Add 250 g/9 oz button mushrooms with the courgettes, step 4.

Pasta ratatouille

10 portions

Add 400 g/14 oz cooked wholemeal pasta, for example short cut macaroni, and 250 g/9 oz diced cooked chicken to the ratatouille during cooking. This makes the ratatouille into a substantial main course dish with 5 g fibre per portion (see Appendix 2).

Fennel ratatouille

10 portions

Fennel will give the ratatouille a Mediterranean flavour which is particularly good served with kebabs. Add 500 g/1 lb 2 oz blanched diced fennel (use a little lemon in the water when blanching the fennel) to the ratatouille at the same time as the aubergines. The feathery green top of the fennel can be used as a garnish instead of the parsley.

Spicy vegetable dishes

Vegetable curry

10 portions

Each portion contains:
709 kJ/168 kcal, 8 g fibre,
4 g fat.

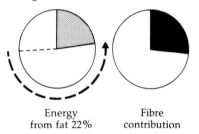

Energy   Fibre
from fat 22%   contribution

30 ml	Polyunsaturated oil	1 fl. oz
200 g	Diced onions	7 oz
500 g	Sliced carrots	1 lb 2 oz
500 g	Diced parsnips	1 lb 2 oz
500 g	Diced potatoes	1 lb 2 oz
250 g	Cauliflower florets	9 oz
250 g	Cut French beans	9 oz
250 g	Cooked haricot beans with cooking liquor	9 oz
30 g	Ground coriander	1 oz
10 g	Ground cumin seeds	1 dspn
5 g	Chilli powder	2 tspns
10 g	Turmeric	1 dspn
2½ g	Mixed spice	1 tspn
1 g	Ground cinnamon	½ tspn
Pinch	Ground cloves	Pinch
Pinch	Ground Methi leaves	Pinch
200 ml	Yogurt	7 fl. oz
5 g	Salt	1 tspn

METHOD

1 Heat the oil in a cooking utensil, add the onions and sweat for a short period.
2 Add the spices and gently cook for 2–3 minutes; take care not to burn the spices.
3 Add the carrots, parsnips, potatoes, yogurt and salt, bring to the boil. At this stage add a little bean cooking liquor to allow the vegetables to cook evenly.

continued overleaf

4 Add the cauliflower and green beans and complete the cooking, allowing excess liquid to evaporate, keeping the vegetables crisp in texture.
5 Add the cooked haricot beans and allow to heat through.

Channa dahl

10 portions

Each portion contains:
847 kJ/200 kcal, 8 g fibre,
6 g fat.

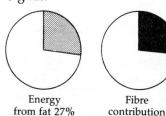

Energy	Fibre
from fat 27%	contribution

30 ml	Polyunsaturated oil	1 fl. oz
120 g	Sliced onions	4 oz
5 g	Ground cumin	2 tspns
5 g	Turmeric	2 tspns
5 g	Ground coriander	2 tspns
2½ g	Chilli powder	1 tspn
15 g	Chopped fresh root ginger	½ oz
500 g	Cooked chick-peas with cooking liquor (or other beans)	1 lb 2 oz
15 g	Chopped fresh coriander leaves	½ oz
5 g	Salt	1 tspn

METHOD

1 Heat oil in a cooking utensil, add onions and sweat for a short period.
2 Add all the spices except the coriander leaves and cook for 2–3 minutes; take care not to burn the spices.
3 Add chick-peas and enough cooking liquor to make a thick mixture.
4 Mix in the chopped coriander leaves and allow to heat through.

Lentil dahl

10 portions

Each portion contains:
795 kJ/188 kcal, 6 g fibre,
4 g fat.

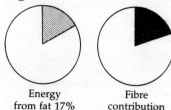

Energy	Fibre
from fat 17%	contribution

500 g	Lentils	1 lb 2 oz
5 g	Salt	1 tspn
30 ml	Polyunsaturated oil	1 fl. oz
120 g	Sliced onions	4 oz
5 g	Chopped garlic	2 tspns
5 g	Turmeric	2 tspns
2½ g	Chilli powder	1 tspn
15 g	Chopped fresh root ginger	½ oz
15 g	Chopped fresh mint leaves	½ oz

METHOD

1 Boil the lentils in water with the salt until soft, drain and place aside.
2 Heat the oil in a cooking utensil, add the onions and garlic and sweat for a short period.
3 Add the turmeric, chilli powder and ginger and cook for 2–3 minutes; take care not to burn the spices.
4 Add the lentils and chopped mint leaves and allow to heat through.

Armenian vegetable stew

10 portions

Each portion contains:
604 kJ/144 kcal, 5 g fibre,
5 g fat.

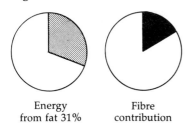

| Energy from fat 31% | Fibre contribution |

10 ml	Olive/sunflower oil	2 tspns
90 g	Sliced onions	3 oz
750 g	Sliced aubergines	1 lb 10 oz
900 g	Sliced courgettes	2 lb
150 g	Sliced peppers	5 oz
250 g	Cut French beans	9 oz
250 g	Cooked haricot beans	9 oz
20 g	Paprika	2 dspns
10 g	Crushed garlic	1 dspn
600 ml	Plain natural yogurt	1 pt 1 fl. oz
240 ml	Eggs	5 eggs
5 g	Salt	1 tspn

METHOD

1 Mix together the yogurt and garlic and place aside.
2 Heat the oil in a cooking utensil, sweat off the onions, then add the aubergines and sweat for 4–5 minutes. Keep the aubergines under-cooked.
3 Arrange the vegetables in layers sprinkled with the paprika in an ovenproof dish.
4 Cover with a tight-fitting lid, then bake in a moderate oven until lightly cooked, 20 minutes approximately.
5 Beat the eggs, then pour evenly across the vegetables and allow to soak through.
6 Return to the oven and bake until set.
7 When cooked, pour over the yogurt and garlic mixture and decorate with a little paprika.

CHAPTER 5

Oven cooking

Many methods of cookery share the common element of using an oven in which to cook food: **shallow poaching (fish), braising, casserole cooking, oven roasting and baking** being prime examples. In some instances there may be an alternative to using an oven when a specialist piece of equipment is available. For example, the shallow poaching of fish and the braising of meat dishes may be done in a Bratt pan (a large tilting cooking container with lid).

An oven may also be used as an alternative to grilling and shallow frying when cooking cuts of fish, butcher meats, game and poultry, for example fish and fish fillets, veal chops, chicken portions, and low-fat sausages. The food is placed on to a tray, very lightly brushed with oil, then cooked in an oven, with development of colour and cooking taking place simultaneously. In addition, many cooks use an oven as a secondary stage to grilling and shallow frying, especially when items are very thick and require long cooking times, for example large steaks and chops; the steak or chop is grilled to develop colour then placed in an oven to complete the cooking.

One important advantage with this type of cooking is that the quantity of fat required for cooking foods can be reduced; **the need for the food to sit in fat or be repeatedly basted with fat can be eliminated**. Also, marinades, which are very low in fat, may be used to keep the food moist and produce a novel taste—a practice to be found in Indian cookery, see tandoori chicken on page 159.

This chapter examines processes concerned with the cooking of fish, poultry, meat and vegetables but does not include braising meat, which is dealt with in the section on stews (page 102), or baking flour products, that is, doughs and pastries, which are dealt with in Chapter 9.

SHALLOW POACHED FISH DISHES

Fish dishes which contain a wide variety of flavoured sauces are usually cooked by shallow poaching. The flavour and important characteristics of all the ingredients are contained within the sauce because the cooking liquor becomes the foundation for the sauce and is not discarded.

Traditionally, shallow poached fish dishes are often prepared with very rich sauces. In some cases the sauce consists of the cooking liquor reduced with cream and thickened with butter; an egg yolk sabayon (egg yolks and wine or water whisked until stiff in a container of hot water) is also added if the sauce is to be browned under a grill.

Avoid these high-fat recipes and follow the procedure given for the following recipes.

Fish Suchet

Example dish: cod Suchet

10 portions

Each portion contains: 692 kJ/164 kcal, 2 g fibre, 2 g fat.

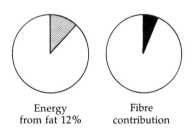

Energy from fat 12%   Fibre contribution

Quantity	Ingredient	Imperial
10 × 150 g	Cod suprêmes (cod fillet portions)	10 × 5 oz
120 ml	Dry white wine	4 fl. oz
300 ml	Fish stock	10½ fl. oz
30 ml	Lemon juice	1 fl. oz
750 ml	Low-fat fish sauce (page 75)	1 pt 6 fl. oz
90 g	Sliced onions	3 oz
90 g	Blanched carrot strips	3 oz
90 g	Blanched celery strips	3 oz
30 g	Blanched turnip strips	1 oz
30 g	Blanched swede strips	1 oz
30 g	Blanched celeriac strips	1 oz
90 g	Leek strips	3 oz
90 g	Sliced cooked green beans	3 oz
To taste	Chopped parsley	To taste

BASIC METHOD FOR SHALLOW POACHED FISH DISHES

1 Place the wine, lemon juice and vegetables into the cooking utensil.
2 Place the fish on top of the ingredients in the cooking utensil, then add the fish stock up to two-thirds the height of the fish approximately.
3 Cover with a piece of oiled paper and lid, bring almost to the boil on top of the stove.
4 Place in a moderate oven to cook, 8–10 minutes.

5 Remove the fish from the pan and keep hot, covered with the oiled cooking paper.
6 Reduce down the cooking liquor and vegetables by two-thirds approximately, then add the fish sauce.
7 Check consistency then add the chopped parsley.
8 Coat the base of the serving dish with a little sauce to avoid the fish sticking, then neatly place on the fish.
9 Coat with the remaining sauce and serve.

Large-scale service
When preparing the fish for large-scale service, the sauce and fish would normally be prepared separately. The sauce would be prepared using the recipe ingredients without the fish from step 6. When required for service the fish is cooked using fish stock. After cooking, the cooking liquor from the fish is reduced down and added to the ready-made sauce.

Fish Véronique

Example dish: fillets of place Véronique

10 portions

Each portion contains:
1455 kJ/344 kcal, 1 g fibre, 10 g fat.

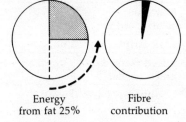

Energy   Fibre
from fat 25%   contribution

10 × 150 g	Plaice fillets	10 × 5 oz
120 ml	Dry white wine	4 fl. oz
300 ml	Fish stock	10½ fl. oz
30 ml	Lemon juice	1 fl. oz
120 g	Finely chopped onions	4 oz
850 ml	Low-fat fish sauce (page 75)	1½ pt
120 g	Grapes (halved and seeded)	4 oz
60 ml	Whipping cream	2 fl. oz

METHOD

1 Prepare as stated in the basic recipe above.
2 *Whisk the cream until stiff, fold into the completed sauce.
3 When serving, dress the grapes on top of the fish, then coat with the sauce. Glaze under a hot salamander (grill) and serve.

*When preparing sauces which are to be glazed under a salamander, a small quantity of whipped cream is a suitable alternative to sabayon. This method can be used for any of the classic fish dishes, for example Fish Bréval.

Seafood mélange

10 portions

Each portion contains:
1388 kJ/280 kcal, 3 g fibre,
5 g fat.

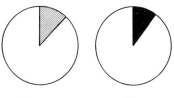

Energy from fat 12%	Fibre contribution	

600 g	Monkfish fillets (in strips)	1 lb 5 oz
150 g	Peeled prawns	5 oz
150 g	Peeled scampi	5 oz
90 g	Shelled cooked mussels	3 oz
200 ml	Fish stock	7 fl. oz
150 g	Diced onions	5 oz
2 g	Crushed garlic	1 large clove
90 g	Diced red peppers	3 oz
150 g	Diced tomatoes	5 oz
150 g	Diced cucumbers	5 oz
150 g	Peeled diced marrow	5 oz
10 g	Minced root ginger	½ oz
5 g	Lemon zest	2 tspns
30 ml	Lemon juice	1 fl. oz
10 ml	Anchovy essence	2 tspns
5 g	Ground coriander	2 tspns
5 g	Ground fennel	2 tspns
5 g	Turmeric	2 tspns
10 g	Curry powder	½ oz
750 ml	Low-fat fish sauce (page 75)	1 pt 6 fl. oz

Accompaniment

*1 kg	Cooked brown rice	2 lb 3 oz

*To attain the approximate quantity of cooked rice, use half the quantity of raw rice to cooked rice.

METHOD

1 Lightly poach the monkfish in the fish stock and lemon juice together with the vegetables and spices.
2 Remove the monkfish and reduce down the cooking liquid and vegetables.
3 Add the monkfish and the shellfish and thoroughly reheat.
4 Add the anchovy essence and fish sauce.
5 Check consistency and serve.

North Sea fish stew

10 portions
Each portion contains:
1748 kJ/414 kcal, 5 g fibre,
10 g fat.

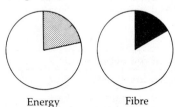

| Energy | Fibre |
| from fat 22% | contribution |

400 g	Monkfish fillets	14 oz
400 g	Cod fillet	14 oz
400 g	Mackerel or herring	14 oz
60 g	Shelled prawns	2 oz
120 g	Mussels, cooked and shelled	4 oz
300 g	Diced onions	10½ oz
300 g	Diced white leeks	10½ oz
5 g	Crushed garlic	2 cloves
450 g	Diced tomatoes	1 lb
200 g	Button mushrooms	7 oz
150 ml	Dry white wine	5 fl. oz
500 ml	Fish stock (to barely cover fish)	18 fl. oz
850 ml	Low-fat fish sauce (page 75)	1½ pt

Accompaniment

*1 kg	Cooked brown rice	2 lb 3 oz

*To attain the approximate quantity of cooked rice, use half the quantity of raw rice to cooked rice.

METHOD

1 Cut the fish into pieces (30 mm/1¼ inches).
2 Lightly poach the fish in the fish stock and white wine.
3 Drain off the cooking liquor into a clean pan, place the fish aside and keep hot.
4 Add the onions, leeks, garlic and tomatoes into the cooking liquor, bring to the boil and reduce down by two-thirds.
5 Add the mushrooms and allow to cook while continuing the reduction of the cooking liquor.
6 Add the shellfish and allow to heat through thoroughly.
7 Add the fish sauce.
8 Carefully add the fish, check temperature and consistency.
9 Serve accompanied with the hot boiled rice.

OTHER FISH DISHES

Fish pie

10 portions

Each portion contains:
1395 kJ/330 kcal, 5 g fibre,
9 g fat.

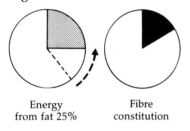

Energy from fat 25%   Fibre constitution

10 × 120 g	Skimmed fish fillets or suprêmes (cod, plaice, sole, etc.)	10 × 4 oz
200 g	Diced onions	7 oz
120 g	Diced red peppers	4 oz
120 g	Cooked sweetcorn kernels	4 oz
200 g	Button mushrooms	7 oz
120 g	Blanched diced green beans	4 oz
30 ml	Lemon juice	1 fl. oz
850 ml	Thin fish sauce (page 75)	1½ pt
To taste	Ground black pepper	To taste
500 g	Short pastry (page 181)	1 lb 2 oz

METHOD

1 Lightly cook the onions, peppers and mushrooms in half the sauce, then quickly cool. Also ensure that the remaining sauce is cool.
2 Coat the base of a suitable cooking utensil with the remaining sauce.
3 Place the fish in portions on the sauce in the cooking utensil, sprinkle over the lemon juice.
4 Place the sweetcorn kernels and beans over the fish, cover with sauce and garnish.
5 Roll out the pastry until it is 4–5 mm/⅛ inch thick and approximately the size of the cooking dish. Cut the pastry to a size slightly larger than the cooking dish using a similar dish as a guide.
6 Lightly dampen the rim of the dish, place on the pastry cover and press down the edges to seal.
7 Notch or mark the edge of the pastry and cut a small hole in the centre.
8 Lightly brush with eggwash, place on a baking tray and bake at 200°C for 20–30 minutes.

Seafood and potato pie

10 portions

Each portion contains:
1371 kJ/325 kcal, 4 g fibre,
9 g fat.

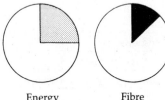

| Energy | Fibre |
| from fat 25% | contribution |

800 g	Fillets of white fish (e.g. cod, plaice, whiting, etc.)	1 lb 12 oz
400 g	Fillets of oily fish (e.g. herring or mackerel)	14 oz
120 g	Shelled prawns	4 oz
120 g	Shelled scampi tails	4 oz
120 g	Mussels, cooked and shelled	4 oz
150 g	Chopped onions	5 oz
2 g	Crushed garlic	1 large clove
120 g	Diced blanched root fennel	4 oz
200 g	Diced tomatoes	7 oz
120 g	Cooked peas	4 oz
Good pinch	Roughly chopped parsley	Good pinch
30 ml	Lemon juice	1 fl. oz
500 ml	Fish stock (to cover)	18 fl. oz
850 ml	Low-fat fish sauce (page 75)	1½ pt
*1 kg	Mashed potatoes	2 lb 3 oz

*Quantity of potatoes may be halved if the pie is to
be served with a potato accompaniment.

METHOD

1 Arrange the fish into portions on the cooking dish, add the onions,
   garlic, lemon juice and fish stock.
2 Lightly poach the fish, drain off the cooking liquor into a clean pan.
3 Reduce down the cooking liquor until thick and concentrated, add to
   the fish sauce.
4 Garnish the fish with shellfish, tomatoes, peas, fennel and parsley.
5 Carefully bind the fish with the sauce, ensuring the fish is not sticking
   to the cooking utensil.
6 Neatly cover with the mashed potatoes.
7 Brush over the surface of the potatoes with a little milk. Bake in
   a moderate oven until golden brown and the filling thoroughly
   reheated.

RICE AND FARINACEOUS DISHES

RIZ PILAFFS AND RISOTTOS

These are basically rice dishes which consist of rice cooked in stock and flavoured with vegetables and herbs. Other ingredients which are added are cheese, spices, nuts, fish and shellfish, poultry and butcher meats.

Riz pilaffs are braised rices of Arabic origin and are sometimes referred to as 'riz pilau'. Riz pilaffs usually contain two parts liquid to one part rice and are cooked in the oven until quite dry in texture.

Risottos from Italian cuisine usually contain three parts liquid to one part rice and are traditionally cooked on top of a stove. They are more moist than riz pilaffs and are ideal as dishes on their own or served as an accompaniment to a dish. Recipes for risottos have been included in this section because of their similarity to riz pilaffs.

Cookery hints

- Always use a good quality hard cooking rice, for example, basmati, patna or piedmont, etc. to avoid a dish with a pudding-like texture.
- The professional cook's basic guide for these recipes is to use volume measures when adding the rice and stock, for example a ladle or jug.

Riz pilaffs:	**1 measure of rice × 2 measures of stock**
Risottos:	**1 measure of rice × 3 measures of stock**

- Always use a fork to stir the rice, especially when cooked; using a spoon or spatula is more likely to burst the rice grains.
- Avoid the traditional practice of forking butter through the cooked rice.
- Brown rice can be used as an alternative to white rice, but the cooking time will need to be approximately doubled (see page 41).

Basic recipe for riz pilaffs

Yield 10 portions

Each portion contains:
1041 kJ/246 kcal, 3 g fibre,
5 g fat.

600 g	*Brown rice	1 lb 5 oz
30 ml	Polyunsaturated oil	1 fl. oz
200 g	Chopped onions	7 oz
1 litre	White stock	1¾ pt
To taste	Ground white pepper	To taste
5 g	Salt	1 tspn

*Long grain white rice may be used as an alternative but will have a lower fibre content.

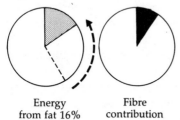

Energy
from fat 16%

Fibre
contribution

METHOD

1 Sweat the onions until soft in a suitable saucepan, 3–4 minutes.
2 Add the rice and stock and bring to the boil.
3 Cover with a piece of lightly oiled greaseproof paper and lid, cook in a moderate oven, 15 minutes approximately.
4 When cooked the stock should have evaporated, leaving the rice grains quite dry and easily separable.
5 Fork through the pepper.

Basic recipe for risottos

(Yield 10 portions)

This is the same as riz pilaff above, with the following amendments:

1 Increase the stock in the recipe to 1½ litres.
2 Add half the quantity of stock and cook slowly over a low heat, covered with a piece of lightly oiled paper and lid. Stir frequently, adding more stock as required. Use a fork near the end of the cooking period to avoid breaking the rice.

Note: When cooked the stock should have evaporated, leaving the rice grains slightly pasty with an attractive moist eating quality.

Braised rice creole

Add 300 g/10 oz sliced mushrooms and 200 g/7 oz diced pimentoes to the onions after sweating and continue cooking for 3–4 minutes before adding the stock. Also add 300 g/10 oz diced tomato flesh to the riz pilaff on completion of cooking and allow to heat through.

Braised rice Indian style

Each portion contains: 1257 kJ/296 kcal, 7 g fibre, 4 g fat.

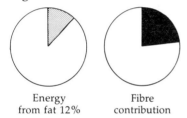

Energy from fat 12%   Fibre contribution

600 g	Long grain rice	1 lb 5 oz
30 ml	Polyunsaturated oil	1 fl. oz
300 g	Diced onions	10½ oz
10 g	Crushed garlic	3 large cloves
10 g	Minced root ginger	½ inch piece
10 g	Ground coriander	1 dspn
10 g	Garam masala	1 dspn
150 g	Sliced mushrooms	5 oz
300 g	Diced green beans	10½ oz
300 g	Cooked mung beans	10½ oz
2½ g	Black mustard seeds	2 tspns
5 g	Chopped parsley	1 dspn
1 litre	Chicken or vegetable stock	1¾ pt
5 g	Salt	1 tspn

METHOD

Follow the basic method, making the following additions:

1   Sweat the onions and garlic in the oil then add the spices and root ginger and cook over a low heat for 1 minute approximately. Avoid burning the spices.
2   Add the sliced mushrooms and cook for a further 2–3 minutes.
3   Add the rice, stock and green beans and cook as stated in the basic recipe.
4   When cooked, fork in the mung beans, mustard seeds and parsley and allow to heat through.

Seafood and salsify risotto

Each portion contains:
1317 kJ/311 kcal, 3 g fibre,
5 g fat.

Energy
from fat 13%

Fibre
contribution

600 g	Long grain rice	1 lb 5 oz
30 ml	Polyunsaturated oil	1 fl. oz
120 g	Chopped onions	4 oz
2½ g	Crushed garlic	1 large clove
60 g	Diced green peppers	2 oz
60 g	Diced red peppers	2 oz
120 g	Sweetcorn kernels	4 oz
60 g	Diced French beans	2 oz
60 g	Cooked peas	2 oz
300 g	Sliced blanched salsify	10½ oz
300 g	Peeled prawns	10½ oz
150 g	Shelled and cooked mussels	5 oz
150 g	Shelled and cooked scampi tails	5 oz
5 g	Turmeric	2 tspns
To taste	Ground black pepper	To taste
1½ litres	Fish stock	2 pt 12 fl. oz
5 g	Salt	1 tspn

METHOD

Follow the basic method, making the following additions:

1 Sweat the onions and garlic in the oil, add the peppers and continue sweating for 2–3 minutes.
2 Add the rice, French beans, turmeric and fish stock and cook as stated.
3 When cooked, fork in the remaining ingredients and allow to heat through thoroughly.

Braised millet

10 portions

600 g	Millet	1 lb 5 oz
30 ml	Polyunsaturated oil	1 fl. oz
200 g	Chopped onions	7 oz
5 g	Crushed garlic	2 cloves
600 ml	White stock	1 pt
To taste	Ground white pepper	To taste
5 g	Salt	1 tspn

METHOD

Prepare the same as braised rice on page 129.

Braised millet Persian style

Each portion contains:
1216 kJ/291 kcal, 4 g fibre,
9 g fat.

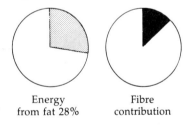

Energy
from fat 28%

Fibre
contribution

METHOD

Prepare the basic braised millet, adding 80 g/3 oz nib almonds when cooking. When cooked, place in the serving dish and garnish with sliced tomatoes (300 g/10 oz) and fresh chopped mint. Accompany with a dish of plain natural yogurt (300 ml/½ pt).

HOTPOTS AND CASSEROLE-STYLE DISHES

Cookery hint

- The quantity of liquid used in these dishes is very important—too little liquid may result in the contents drying out and burning, too much liquid will result in the hotpots or vegetables becoming soggy and waterlogged. *The rule is to add liquid until it is just below the top of the main items; less liquid should also be added to watery vegetables which do not have a long cooking time.*

Beef carbonnade with vegetables

10 portions
Each portion contains:
813 kJ/193 kcal, 3 g fibre,
8 g fat.

| Energy | Fibre |
| from fat 35% | contribution |

1 kg	Sliced lean beef	2 lb 3 oz
30 ml	Polyunsaturated oil	1 fl. oz
400 g	Sliced onions	14 oz
200 g	Sliced carrots	7 oz
200 g	Sliced celery	7 oz
200 g	Sliced leeks	7 oz
200 g	Sliced green peppers	7 oz
600 ml	Beer	1 pt
600 ml	Brown stock	1 pt
To taste	Ground black pepper	To taste
	Chopped parsley	
5 g	Salt	1 tspn

METHOD

1 Shallow fry the onions in the oil until lightly coloured.
2 Place the meat and vegetables in layers in a casserole. Complete the top with a layer of the fried onions.
3 Add the stock and beer until just barely covering the meat (but not the final layer of onions).
4 Cover with a tight fitting lid then cook in a moderate oven—1½ hours approximately.
5 When cooked, remove any surface fat and clean round the sides of the casserole. Sprinkle with the parsley and serve.

Farmhouse casserole

10 portions

Each portion contains:
890 kJ/212 kcal, 4 g fibre,
7 g fat.

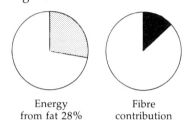

Energy
from fat 28%

Fibre
contribution

10	Skinned chicken quarters	10
200 g	Peeled button onions	7 oz
200 g	Diced carrots	7 oz
200 g	Diced turnips	7 oz
200 g	Sliced celery	7 oz
200 g	Sliced leeks	7 oz
120 g	Sliced cabbage	4 oz
400g	Sliced potatoes	14 oz
1½ litres	White stock	2 pt 12 fl. oz
To taste	Ground black pepper	To taste
5 g	Salt	1 tspn

METHOD

1   Line the casserole with a layer of the sliced potatoes, place half the vegetables on top of the potatoes.
2   Place the chicken quarters on top of the vegetables in the casserole and arrange the remaining vegetable garnish on top.
3   Add the stock to just below the top of the chicken pieces.
4   Cover with the rest of the potatoes, neatly overlapping in rows.
5   Lightly brush over the surface of the potatoes with a little oil, cook in a moderate oven (190–200°C/375–400°F) until lightly coloured.
6   Reduce the temperature to 175°C/350°F approximately and cook slowly, occasionally pressing down with a fish slice. Cooking time: 1 hour 15 minutes approximately.

Vegetable hotpot

10 portions

Each portion contains:
697 kJ/164 kcal, 6 g fibre,
4 g fat.

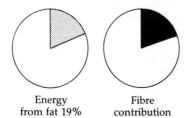

Energy	Fibre
from fat 19%	contribution

400 g	Sliced onions	14 oz
400 g	Sliced potatoes	14 oz
200 g	Diced carrots (large dice)	7 oz
200 g	Diced swedes (large dice)	7 oz
120 g	Diced red peppers (as above)	4 oz
120 g	Diced green peppers (as above)	4 oz
200 g	Diced celery (as above)	7 oz
200 g	Diced leeks (as above)	7 oz
200 g	Diced peeled marrow (as above)	7 oz
120 g	Sweetcorn kernels	4 oz
120 g	Lightly cooked butter beans	4 oz
120 g	Lightly cooked lentils	4 oz
120 g	Lightly cooked brown rice	4 oz
60 g	Shelled peanuts (unsalted)	2 oz
200 g	Sliced tomatoes	7 oz
10 g	Crushed garlic	4 cloves
2 g	Oregano	2 tspns
1½ litres	Vegetable stock	2 pt 12 fl. oz
To taste	Ground black pepper	To taste
60 g	Tomato purée	2 oz
5 g	Salt	1 tspn

METHOD

1 Place a layer of sliced potatoes on the base of the casserole or cooking utensil.
2 Add a layer of sliced onions.
3 Arrange the vegetables in layers up to the top of the cooking utensil, reserving a final layer of sliced potatoes for the top.

Note: It is a good idea to alternate the layers of butter beans, lentils and brown rice with layers of fresh vegetables. The oregano and nuts are sprinkled over the vegetables at this stage.

4 Add the crushed garlic and tomato purée to the stock and mix together.

5 Add the stock until two-thirds the height of the vegetables, then cover with the remaining potatoes in overlapping rows across the top.

6 Lightly brush over the surface of the potatoes with oil, cook in a moderate oven (200°C/400°F) without a lid. Occasionally press down the surface with a fish slice to produce a firm texture. Cooking time: 1½ hours approximately.

Vegetable and cheese hotpot

10 portions

Each portion contains:
875 kJ/208 kcal, 7 g fibre,
5 g fat.

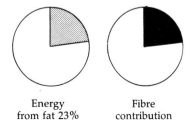

Energy   Fibre
from fat 23%   contribution

METHOD

Prepare the same as vegetable hotpot, adding a topping of grated low-fat cheese and wholemeal breadcrumbs (120 g/4 oz cheese and 60 g/2 oz breadcrumbs) to the hotpot when cooked. Allow to gratinate in the oven or under a hot grill prior to service.

Alubias

10 portions

Each portion contains;
1397 kJ/332 kcal, 17 g fibre,
11 g fat.

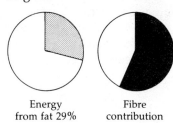

| Energy | Fibre |
| from fat 29% | contribution |

This is a Spanish casserole based on large white haricot beans, flavoured with the bones of the famous raw smoked hams which hang from the ceilings of bars throughout Spain. Any variety of white or green beans can be used and, if available, a small piece of paprika salami sliced into the casserole will give it an authentic taste. Serve alubias as a main course with green salad.

600 g	Raw haricot beans	1 lb 5 oz
30 ml	Corn oil	1 fl. oz
300 g	Sliced onions	10½ oz
*1 litre	Tomato juice	1¾ pt
*800 g	Diced fresh tomatoes	1 lb 12 oz
300 g	Diced lean raw ham	10½ oz
600 g	Diced green peppers	1 lb 5 oz
120 g	Sliced paprika salami	4 oz
5 g	Oregano	1 dspn
150 g	Tomato purée	5 oz
500 ml	Stock or water	18 fl. oz
To taste	Chopped chives (or parsley)	To taste

*850 g tin of tomatoes (A2½) can be substituted for the fresh tomatoes and the tomato juice. These ingredients may also have added sugar. *Check the label.*

METHOD

1 Soak the beans overnight in cold water. Drain off the water when ready for cooking.
2 Heat the oil in a saucepan, shallow fry the onions until a rich brown colour.
3 Combine all the ingredients except the chopped chives in a brasing pan or casserole, then add the tomato juice.
4 Mix in the stock until barely covered.
5 Tightly cover the cooking utensil, cook in an oven at 180°C/350°F for 2 hours approximately. The alubias is cooked when the beans are tender and most of the cooking liquor has been absorbed. *Important:* Take care that the alubias does not dry out during cooking.
6 Check the seasoning, the use of ham may mean that salt is not required.

Note: If there is excess cooking liquor, the dish may be thickened with a flour and water paste at the end of the cooking period; see white sauce on page 73.

Baked filled potatoes

10 portions

Each portion contains:
1183 kJ/278 kcal, 5 g fibre, 4 g fat.

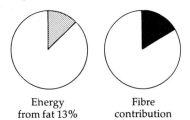

Energy from fat 13%	Fibre contribution

Baked potatoes split open and filled with a mixture of meat and vegetables make filling meals and have become popular in fast food restaurants.

10 × 200 g	Potatoes	10 × 7 oz
600 ml	Low-fat white sauce	1 pt
250 g	Diced cooked chicken or alternative meat	9 oz
250 g	Selected vegetables: diced onions, red and green peppers, diced mushrooms, sweetcorn and peas	9 oz
5 g	Salt	1 tspn

METHOD

1 Wash the potatoes then bake until tender.
2 Add the onions and peppers to the sauce and cook for 3–4 minutes (to lightly soften the vegetables).
3 Add the chicken, sweetcorn, mushrooms and peas and simmer for 2 minutes approximately.
4 Split the hot baked potatoes and place on a serving dish.
5 Fill with the garnished sauce and serve with a side salad.

Note: If desired, 60 g/2 oz low-fat cheddar-type cheese may be added to the vegetables and sauce after step 3.

Cheese and vegetable potatoes

Replace the chicken with a low-fat curd cheese, which should be blended through the sauce after step 3. Also add 1 teaspoon of mustard powder which has been mixed with a little cold water.

Alternative fillings

Almost any diced vegetables can be used in the sauce. Simply blanch the vegetables which require cooking and add to the sauce. Examples are leeks, celery, carrots, celeriac and courgettes. Diced tomatoes should be added at the last minute to avoid breaking up.

OVEN ROASTING

The expected reaction to this method of cooking is to classify it as high in fat. The reason for this is the fatty nature of the foods to be cooked, together with the use of fat as a cooking medium. However, *a considerable reduction in fat content can be achieved by carefully selecting the article to be cooked and amending traditional practice.*

ROASTING BUTCHER MEATS, POULTRY AND GAME

Cookery hints

- Always select lean joints, for example good quality topside and rump beef. This also applies to poultry, which can have a very fatty skin. See commodities on page 48.
- Trim off all excess fat prior to cooking.
- Raise the joint or bird off the roasting tray and out of any drippings. This may be done using a trivet or large pieces of carrot, onion and celery.
- Lightly brush the joint with polyunsaturated oil (not dripping or lard) prior to cooking. Never baste with the drippings.
- Use low oven temperatures when cooking; high temperatures mean more shrinkage and weight loss and less juiciness. *This is very important when cooking lean joints to ensure a tender, juicy product.*
- Remove excess fat from the joint when carving and portioning.
- Thoroughly skim off surface fat from the gravy, then remove all trace fat by floating a clean piece of kitchen paper or dish paper on the surface.
- Do not serve accompaniments which are very high in fat, for example, traditional Yorkshire pudding. Cook Yorkshire puddings in non-stick trays which have only been polished with oil.

ROASTING VEGETABLES (potatoes and parsnips)

Cookery hint

- Use the following method which incorporates the simple technique of brushing the vegetable and roasting tray with oil instead of tossing in fat, which has a frying effect. It is important not to turn the vegetable during cooking.

Roast potatoes

10 portions

Each portion contains:
632 kJ/148 kcal, 3 g fibre,
2 g fat.

1½ kg	Peeled potatoes	3 lb 6 oz
20 ml	Polyunsaturated oil	2 dspns

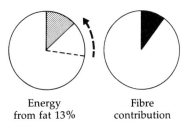

Energy
from fat 13%

Fibre
contribution

METHOD

1 Cut the potatoes into large even pieces (two per portion).
2 Brush the base of the roasting tray with the oil.
3 *Add the potatoes to the tray and brush with the remaining oil.
4 Cook in a hot oven until crisp and golden brown.

*Par-boiling the potatoes for a very short period will reduce the likelihood of the potatoes sticking to the roasting tray, especially when using aluminium trays. The crisp surface should also last for a longer period prior to service.

Yorkshire puddings

10 portions (small)

Each portion contains:
262 kJ/62 kcal, <1 g fibre,
1 g fat.

60 g	Wholemeal flour	2 oz
60 g	Plain flour	2 oz
50 ml	Egg	1 egg
300 ml	Skimmed milk	½ pt
Pinch	Pepper	Pinch
2½ g	Salt	½ tspn
5 ml	Polyunsaturated oil	1 tspn

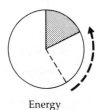

Energy
from fat 18%

METHOD

1 Mix together the two flours and the salt and pepper.
2 Add the egg and half the milk, whisk together until smooth.
3 Add the remaining milk in stages, whisking between each addition, until a smooth batter is obtained.
4 Allow the batter to rest for 1 hour approximately before use.
5 Place the Yorkshire pudding tins on a tray, lightly brush with the oil.
6 Place the tins in a moderate oven (200°C/400°F) and allow to become very hot, 8–10 minutes in the oven.
7 Pour the batter into the hot moulds then replace into the oven.
8 Allow to cook for 30 minutes approximately (until risen and set).
9 Turn the puddings upside down to remove any traces of fat from the centres and continue cooking until crisp, a further 10 minutes approximately.

CHAPTER 6

Frying

SHALLOW FRYING

Shallow frying is a very general method of cooking which may be applied to a wide range of commodities, for example fish, meat, poultry, game, fruit and vegetables. One major disadvantage with this form of cooking is that many foods which are fried not only get coated with fat but also absorb the fat during cooking. Therefore the golden rule when shallow frying is to use very little fat (only enough fat to stop the food sticking to the pan) and where possible dry fry on a non-stick surface. Unfortunately, this can create difficulties in a commercial situation with aluminium pans and when cooking thick cuts of meat. However, in many instances *part dry frying followed by oven cooking is a good solution. Also, many items may be grilled rather than shallow fried; this being a better method for reducing the fat content of meat, poultry, game and made up dishes.*

Since the preference is to grill (or dry fry then oven cook) as many commodities as possible, this chapter is primarily concerned with shallow fried dishes prepared with a sauce and stir fried dishes.

SAUTÉS OF BUTCHER MEATS, POULTRY, GAME AND OFFAL

This method of cooking produces the finest meat, poultry, game and offal dishes which include a sauce.

There are two important stages when producing these dishes.

1 The main commodity is shallow fried until cooked. Therefore only commodities which are good quality should be used, that is, ones which can be cooked tender by shallow frying.
2 When cooked, the commodity is removed from the pan (and kept hot), then any fat is carefully decanted off, leaving the sediment in the pan. For small numbers and à la carte service the accompanying sauce is then made in the cooking utensil, capturing the aroma and flavour lost during cooking. To complete the dish the commodity and sauce are brought together for service.

For large numbers the production of the sauce in the cooking utensil may be time-consuming, therefore the sauce can be prepared in advance. When the main commodity and any cooking fat have been removed, the cooking utensil is swilled with a little stock or water. This is then boiled for a short period to absorb the flavour, then added to the sauce.

Cookery hints

- Items which are difficult to cook, for example thick pieces of chicken, may be part fried then cooked in an oven.
- When combining the main commodity and sauce avoid further cooking as this will toughen the flesh or produce a taste and texture similar to a stew.
- Some cooks complete the cooking of chicken in the recipe sauce because it must be cooked through to avoid food poisoning. When this is done the previous point on the taste and texture of the cooked flesh must be considered.

Hungarian-style chicken with butter beans

10 *portions*

Each portion contains:
2214 kJ/506 kcal, 5 g fibre, 14 g fat.

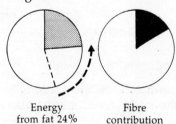

Energy from fat 24% Fibre contribution

30 ml	Polyunsaturated oil	1 fl. oz
10 × 225 g	Chicken quarters	10 × 8 oz
Sauce		
150 g	Chopped onions	5 oz
60 g	Paprika	2 oz
150 ml	White wine	5 fl. oz
300 g	Diced tomato flesh	10½ oz
300 g	Cooked butter beans	10½ oz
700 ml	Low-fat chicken sauce (page 75)	1¼ pt
Accompaniment		
1 kg	Boiled brown rice	2 lb 3 oz

METHOD

Preparing the sauce

1 Place the onions, white wine and paprika in a saucepan and cook until the white wine has reduced down by two-thirds.

2 Add the tomatoes and cook for a short period, 2–3 minutes.
3 Add the chicken sauce, bring to the boil, add the beans and thoroughly reheat.
4 Check consistency and place aside for service.

Cooking the chicken
5 Shallow fry the chicken until coloured on both sides then cover with a lid and place in a moderate oven.
6 When cooked, remove the chicken then decant off any fat from the cooking utensil.
7 Place the cooking utensil back on the heat, add a little stock or water (120 ml/4 fl. oz approximately) and bring to the boil.
8 Simmer for 1 minute then add to the sauce.
9 Place the cooked chicken on a serving dish and coat with the hot sauce. Alternatively, place the chicken and sauce in a container ready for service.
10 Serve the boiled brown rice separately.

Chicken chasseur

10 portions

Each portion contains:
1172 kJ/279 kcal, 1.5 g fibre, 11 g fat.

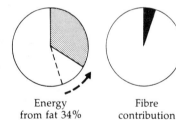

Energy from fat 34%   Fibre contribution

20 ml	Polyunsaturated oil	2 dspns
10 × 225 g	Chicken quarters	10 × 8 oz
Sauce		
Pinch	Chopped tarragon	Pinch
150 ml	White wine	5 fl. oz
30 ml	Brandy	1 fl. oz
80 g	Chopped onions	3 oz
200 g	Sliced mushrooms	7 oz
200 g	Diced tomato flesh	7 oz
700 ml	Low-fat brown sauce (page 76)	1¼ pt
5 g	Chopped parsley	1 dspn

METHOD

Preparing the sauce
1 Place the white wine, brandy, tarragon and onions in a saucepan and cook until the wine has reduced down by two-thirds.
2 Add the sliced mushrooms and cook until the mushrooms soften, 4–5 minutes.

continued overleaf

3   Add the tomatoes and cook for a short period, 2–3 minutes.
4   Add the sauce and bring to the boil.
5   Add the parsley, check consistency.
6   Cook and dish the chicken as above, that is, Hungarian-style chicken.

Tournedos with mead and vegetable sauce

10 portions

Each portion contains:
1198 kJ/283 kcal, 3 g fibre,
9 g fat.

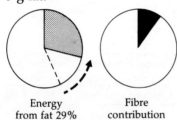

Energy
from fat 29%

Fibre
contribution

30 ml	Polyunsaturated oil	1 fl. oz
10 × 170g	Tournedos (medium size fillet steaks)	10 × 6 oz
Sauce		
150 ml	Mead	5 fl. oz
10 ml	Lemon juice	Squeeze
700 ml	Low-fat brown sauce (page 76)	1¼ pt
150 g	Shredded onions	5 oz
150 g	Shredded leeks	5 oz
150 g	Blanched carrot strips	5 oz
150 g	Blanched celery strips	5 oz
150 g	Cooked lima beans	5 oz
To taste	Ground black pepper	To taste

METHOD

A   Preparing the sauce
1   Place the mead, lemon juice, shredded onions and leeks into a sauce-pan and cook until the mead has reduced down by two-thirds.
2   Add the sauce and remaining vegetables and beans and bring to the boil. Allow the vegetables to be thoroughly reheated.
3   Check consistency and season with the pepper.

B   Cooking the steaks
4   Heat the oil in a suitable pan, shallow fry the steaks to the desired degree of cooking.
5   Remove the steaks and keep hot.
6   Decant off any fat from the cooking utensil, place back on the heat, then add a little stock or water.
7   Simmer for 1 minute then add to the sauce.
8   Place the steaks on the serving dish, coat with the sauce and serve immediately.

Escalopes of veal with succotash and green grape sauce

10 portions

Each portion contains:
1241 kJ/295 kcal, 7 g fibre,
8 g fat.

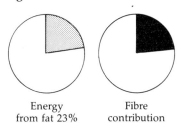

Energy from fat 23%	Fibre contribution

30 ml	Polyunsaturated oil	1 fl. oz
10×100 g	Veal escalopes	10 × 4 oz
	Flour for coating escalopes	

Sauce

80 g	Finely chopped onions	3 oz
140 g	Dry white wine	5 fl. oz
300 ml	Low-fat natural yogurt	½ pt
400 ml	Low-fat veal or chicken sauce (page 75)	14 fl. oz
150 g	Green grapes (halved and pipped)	5 oz

Succotash

500 g	Hot sweetcorn kernels	1 lb 2 oz
400 g	Hot lima beans	14 oz
120 ml	Low-fat chicken sauce (to bind the vegetables	4 fl. oz

METHOD

A *Preparing the sauce*
1 Place the onions and white wine in a saucepan and cook until the wine has reduced down by two-thirds.
2 Add the sauce and bring to the boil.
3 Add the yogurt and green grapes and allow to become hot.

B *Cooking the escalopes*
4 Heat the oil in a pan suitable for shallow frying.
5 Pass the escalopes through the flour then shallow fry until cooked without developing colour.
6 Remove from the cooking utensil and keep hot.
7 Decant off any fat from the cooking utensil, place back on the heat then add a little stock or water.
8 Simmer for 1 minute then add to the sauce.

C *Preparing the succotash*
9 Bind the hot vegetables with the sauce.
10 Arrange the succotash as a border around the serving dish, place the escalopes in the centre, coat with the sauce.

Medaillons of venison with spicy red wine sauce and water chestnuts

10 portions

Each portion contains:
1416 kJ/337 kcal, 1 g fibre,
12 g fat.

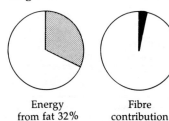

Energy from fat 32%	Fibre contribution	

30 ml	Vegetable oil	1 fl. oz
30 × 40 g	Thin slices of venison (good quality)	30 × 1½ oz
	Flour for coating	

Sauce

80 g	Chopped onions	3 oz
5 g	Crushed garlic	2 cloves
5 g	Grated lemon zest	2 tspns
20 g	Grated root ginger	1 inch piece
2½ g	Allspice	1 tspn
20 g	Grainy-type mustard	1 dspn
140 ml	Red wine	5 fl. oz
700 ml	Low-fat brown sauce (page 76)	1¼ pt
300 g	Sliced water chestnuts (tinned)	10½ oz
To taste	Ground black pepper	To taste

METHOD

A *Preparing the sauce*
1 Place the onions, garlic, lemon zest, grated root giner, allspice, mustard and red wine into a saucepan.
2 Cook until the wine has reduced down by two-thirds.
3 Add the brown sauce and water chestnuts and bring to the boil.
4 Check consistency and season with the pepper.

B *Cooking the médaillons*
5 Cook in the same manner as escalopes above, but allow to colour.
6 Swill the cooking utensil with stock or water as in the previous recipes and add to the sauce.
7 Place the médaillons on a serving dish and coat with the sauce.

Lambs' kidneys with Madeira sauce, rice and Chinese vegetables

10 portions

Each portion contains:
731 kJ/174 kcal, 2 g fibre,
7 g fat.

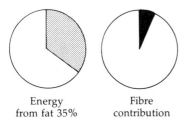

Energy from fat 35%   Fibre contribution

30 ml	Polyunsaturated oil	1 fl. oz
20	Lambs' kidneys	20

Sauce

80 g	Chopped onions	3 oz
80 ml	Madeira wine	3 fl. oz
700 ml	Low-fat brown sauce (page 76)	1¼ pt
To taste	Cayenne pepper	To taste

Chinese vegetables

150 g	Beansprouts	5 oz
150 g	Bamboo shoots (tinned)	5 oz
150 g	Sliced water chestnuts (tinned)	5 oz
150 g	Blanched carrot strips	5 oz

Rice

1 kg	Braised rice (page 128)	2 lb 3 oz

(not included in the nutrient analysis)

METHOD

A *Preparing the sauce*
1 Place the onions and Madeira wine into a saucepan and cook until the wine has reduced down by two-thirds.
2 Add the brown sauce and bring to the boil.
3 Check consistency and season with the cayenne pepper.

B *Preparing the Chinese vegetables*
4 Place the beansprouts into a little boiling water, add the remaining vegetables and thoroughly reheat.

C *Preparing and cooking the kidneys*
5 Remove the skin from the kidneys then cut into halves.
6 Trim off the centre fat then cut at a slant into neat pieces.
7 Heat the oil, quickly shallow fry the kidneys, keeping them a little underdone.
8 Remove the kidneys, decant off the fat, swill the pan with a little stock or water as in the previous recipes and add to the sauce.
9 Add the cooked kidneys to the sauce and check temperature.
10 Serve the kidneys and sauce on a bed of the braised rice then neatly arrange the Chinese vegetables over the top.

STIR FRY DISHES

Stir frying is a popular method of preparing meat and poultry dishes and is the Chinese equivalent of the cooking method 'sauter', already dealt with in this chapter. Unfortunately stir frys usually contain more fat or oil than is desirable. However, the character and taste of these dishes can still be enjoyed when less fat is used. The following recipes contain approximately half the fat of conventional stir fry recipes.

Stir fried chicken with mushrooms, bamboo shoots and bean sprouts

10 portions

Each portion contains:
1782 kJ/425 kcal, 1.5 g fibre, 10 g fat.

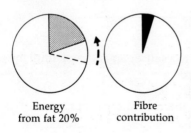

Energy from fat 20%

Fibre contribution

60 ml	Polyunsaturated oil	2 fl. oz
1 kg	Boned sliced chicken	2 lb 3 oz
120 g	Sliced onions	4 oz
5 g	Crushed garlic	2 cloves
20 g	Grated root ginger	1 inch piece
60 g	Chopped spring onions	2 oz
60 g	Blanched strips of celery	2 oz
250 g	Sliced mushrooms	9 oz
120 g	Sliced bamboo shoots (tinned)	4 oz
120 g	Bean sprouts	4 oz
60 ml	Light soy sauce	2 fl. oz
30 ml	Lemon juice	1 fl. oz
30 g	Honey	1 oz
400 ml	Chicken stock (to cover)	14 fl. oz
30 g	Cornflour	1 oz

Accompaniment

1 kg	*Cooked long grain rice	2 lb 3 oz

*Brown rice will increase fibre to 4 g per portion

METHOD

1 Heat approximately one-third of the oil in the cooking utensil.
2 Add the onions, spring onions, garlic and celery, and stir fry for 30 seconds.

3 Add the mushrooms and stir fry for a further short period (until the mushrooms are lightly cooked).
4 Remove the vegetables from the cooking utensil and keep hot.
5 Add the remaining oil to the cooking utensil and place over the heat.
6 Add the chicken, quickly stir fry until cooked.
7 Add the honey, root ginger, lemon juice and two-thirds of the chicken stock.
8 Bring to the boil then lightly thicken with the cornflour diluted in a little cold water.
9 Simmer for 30 seconds to cook the cornflour then add the bamboo shoots, bean sprouts and stir fried vegetables.
10 Heat thoroughly, keeping the vegetables crisp in texture.
11 Adjust the consistency with the remaining stock if required.
12 Accompany with the long grain rice.

Note: For large-scale recipes it may be desirable to use two cooking utensils, both the vegetables and the chicken being cooked at the same time, then later mixed together.

Stir fried chicken with green peppers and mung beans

10 portions
Each portion contains:
1814 kJ/429 kcal, 4 g fibre, 10 g fat.

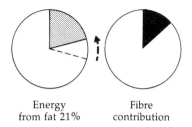

Energy from fat 21%	Fibre contribution	

60 ml	Polyunsaturated oil	2 fl. oz
1 kg	Boned sliced chicken	2 lb 3 oz
120 g	Sliced onions	4 oz
10 g	Crushed garlic	4 cloves
10 g	Grated root ginger	½ inch piece
60 g	Chopped spring onions	2 oz
150 g	Sliced green peppers	5 oz
120 g	Cooked mung beans	4 oz
30 ml	Lemon juice	1 fl. oz
60 ml	Light soy sauce	2 fl. oz
30 g	Honey	1 oz
400 ml	Chicken stock (to cover)	14 fl. oz
30 g	Cornflour	1 oz

Accompaniment

1 kg	Cooked long grain rice	2 lb 3 oz

See overleaf for method

METHOD

Proceed in the same manner as the stir fried recipe on page 148.

Stir fry the vegetables (except the mung beans) in approximately one-third of the oil. Remove the vegetables, stir fry the chicken in the remaining oil. When cooked, add the root ginger, lemon juice, soy sauce, honey and chicken stock and bring to the boil. Thicken with the cornflour diluted in cold water then add the mung beans and stir fried vegetables. Check temperature and consistency. Serve accompanied with the boiled rice.

Chicken and vegetable khoreshe (Iranian)

10 portions

Each portion contains: 2284 kJ/542 kcal, 5 g fibre, 16 g fat.

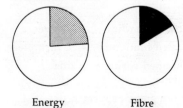

Energy from fat 24%	Fibre contribution

30 ml	Polyunsaturated oil	1 fl. oz
150 g	Diced onions	5 oz
10 g	Crushed garlic	4 cloves
170 g	Diced green peppers	6 oz
300 g	Diced courgettes	10½ oz
150 g	Diced blanched carrots	5 oz
150 g	Diced blanched celery	5 oz
300 g	Diced tomatoes	10½ oz
60 g	Walnut pieces	2 oz
2½ g	Ground cinnamon	1 tspn
To taste	Ground nutmeg	To taste
To taste	Paprika	To taste
To taste	Turmeric	To taste
To taste	Ground black pepper	To taste
5 g	Grated lemon zest	2 tspns
30 ml	Lemon juice	1 fl. oz
400 ml	Chicken stock (to cover)	14 fl. oz
20 g	Cornflour (optional)	1 dspn
1 kg	Cooked skinned sliced chicken	2 lb 3 oz

Garnish

10 ml	Vegetable oil	2 tspns
150 g	Sliced onions	5 oz
5 g	Crushed garlic	2 cloves
To taste	Chopped thyme	To taste

Accompaniment

| 1 kg | Cooked brown rice | 2 lb 3 oz |

METHOD

1 Shallow fry the onions and garlic in a suitable cooking utensil until softened.
2 Add the spices and lemon zest and continue cooking over a low heat for 3–4 minutes.
3 Add the green peppers, courgettes, tomatoes, lemon juice and two-thirds of the stock and bring to the boil.
4 Simmer until the vegetables are almost cooked.
5 Add the chicken, carrots, celery and walnuts and thoroughly reheat. Adjust the consistency with additional stock if required.
6 Optional: thicken with the cornflour diluted in a little cold water.
7 Check consistency and season with the pepper.
8 Prepare the garnish: shallow fry the onions and garlic in the oil then add the thyme.
9 Serve the khoreshe on a bed of boiled rice with the garnish neatly placed across the top.

DEEP FRYING

This is a method of cooking which directly involves fat as a principal ingredient in all recipes. Therefore it is recommended that the consumption of these dishes must be kept to a minimum to be in accordance with the dietary recommendations stated in Chapter 1.

Cookery hints

The following hints are concerned with reducing the amount of fat absorbed by foods during frying, care of frying medium, and safety.

- Use polyunsaturated oil for deep frying, never lard or dripping.
- Avoid overheating the frying oil. Exposing the oil to excessive temperatures causes rapid deterioration.
- Always use the correct frying temperature and check the accuracy of the thermostat at regular intervals.
- Dry wet foods thoroughly before frying.
- Avoid salting fried foods.
- Coat foods properly prior to deep frying.
- Never overload the fryer, especially when frying high water-containing foods. This is not only dangerous but will also reduce the correct frying temperature, resulting in more fat being absorbed into the food.

- Always use the correct quantity of frying oil. This is sometimes indicated on the side of the frying vessel and is approximately one-half to two-thirds the volume of the frying reservoir.
- Fry as quickly as possible because *low frying temperatures mean longer cooking times and increased fat absorption*.
- Some foods can be part-cooked then only finished by flash frying in hot oil. Chips cut thickly may be boiled until just cooked then well drained (and dried enough to be safe to fry) and deep fried. However, this application for chips may be unrealistic in a commercial situation.
- Flash frying in hot fat for a very short period followed by oven cooking (preferably convection oven cooking) should be practised where possible, for example deep fried meat and poultry dishes, savoury croquettes and Scotch eggs, etc.
- Always drain the foods thoroughly after frying.
- Use absorbent kitchen paper to remove surface fat after draining.
- Never deep fry food when an alternative method of cooking may be used, for example low-fat sausages which should be grilled and bread garnishes for soups and main course dishes which can be oven cooked.

CHAPTER 7

Grilling

Grilling offers the possibility of two main cooking styles: cooking over or under the heat source. This method of cookery should always be selected in preference to shallow frying, provided the food being cooked is not left in contact with fat drippings—cooking taking place on a wire grill tray or grill bars.

This chapter has been designed to include dishes which have a high vegetable content and also illustrates the use of spices and low-fat marinades.

Cookery hints

- When grilling fish, coat with flour, wholemeal flour, cereals or wholemeal breadcrumbs to develop an attractive brown colour and crisp texture.
- Lightly brush the food with polyunsaturated oil to prevent the surface drying out and becoming hard. Use low-fat marinades for this purpose where appropriate.
- Keep an eye on the speed of cooking; too high a temperature and too long a cooking period result in rapid deterioration of many grilled dishes, particularly vegetable products.
- Speed of cooking is important to achieve the desired brown colour and the correct degree of cooking; the thicker the item, the slower the speed of cooking. Also, cook quickly thin items and items to be kept under-done.
- Avoid excessive cooking temperatures as these increase shrinkage and weight loss.
- Always cook through items such as chicken and pork, and never serve underdone, as there is a risk of food poisoning.

BURGER-STYLE DISHES

Vegetable burgers

10 × 100 g/3½ oz portions
approximately

Each portion contains:
729 kJ/174 kcal, 5 g fibre,
3 g fat.

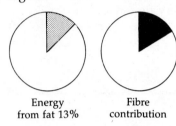

Energy   Fibre
from fat 13%   contribution

120 g	Chopped onions	4 oz
5 g	Crushed garlic	2 cloves
60 g	Chopped chives	2 oz
120 g	Grated carrots	4 oz
120 g	Chopped celery	4 oz
120 g	Sweetcorn kernels	4 oz
120 g	Cooked millet seeds	4 oz
120 g	Cooked mung beans	4 oz
120 g	Cooked lentils	4 oz
60 g	Porridge oats	2 oz
5 g	Fresh chopped parsley	1 dspn
120 g	Wholemeal breadcrumbs	4 oz
20g	Tomato purée	½ oz
10 ml	Lemon juice	2 tspns
10 g	Vegetable extract (Vecon)	1 dspn
To taste	Ground black pepper	To taste
To taste	Ground chilli pepper	To taste
To taste	Ground coriander (optional)	To taste
To taste	Ground cumin (optional)	To taste

Thickening paste

200 ml	Water	7 fl. oz
60 g	Besan flour	2 fl. oz
10 ml	Polyunsaturated oil (for cooking)	2 tspns

METHOD

1 Prepare the thickening paste by boiling half the water, then whisking in the remaining flour and cold water blended to a paste. Cook until very thick then allow to cool. *Note:* Cover the thickening paste or sprinkle with a little cold water prior to cooling to reduce the risk of skin forming.
2 Mix all the ingredients together including the thickening paste, which will bind the mixture.
3 Divide the mixture into portions then shape into burgers using a little flour to avoid sticking.
4 Place the burgers on a grilling try and lightly brush over the surfaces with oil.
5 Grill on both sides until cooked and lightly coloured.
6 Serve accompanied with a tomato, devil or mustard sauce.

Beef and vegetable galettes

(extended burgers)

10 × 120 g portions

Each portion contains: 631 kJ/150 kcal, 4 g fibre, 4 g fat.

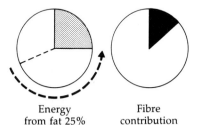

Energy from fat 25%   Fibre contribution

120 g	Chopped onions	4 oz
2½ g	Crushed garlic	2 cloves
120 g	Grated carrots	4 oz
120 g	Blanched diced celery	4 oz
120 g	Cooked lentils	4 oz
120 g	Cooked mung beans	4 oz
120 g	Wholemeal breadcrumbs	4 oz
600 g	Lean minced beef	1 lb 5 oz
60 ml (2 eggs)	Egg whites	2 fl. oz
To taste	Ground black pepper	To taste
5 g	Salt	1 tspn
10 ml	Polyunsaturated oil (for cooking)	2 tspns

METHOD

1 Thoroughly mix together all the ingredients.
2 Cook and serve as vegetable burgers above.

Savoury fish cakes

10 × 120 g portions

Each portion contains:
440 kJ/104 kcal, 2 g fibre,
1.5 g fat.

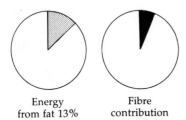

| Energy from fat 13% | Fibre contribution |

150 g	Chopped onions	5 oz
150 g	Chopped leeks	5 oz
120 g	Cooked puréed butter beans	4 oz
300 g	Dry mashed potatoes	10½ oz
500 g	Flaked cooked fish	1 lb 2 oz
5 g	Chopped parsley	1 dspn
To taste	Ground black pepper	To taste
60 ml	Egg whites	2 fl. oz
5 g	Salt	1 tspn

Coating

	Flour	
	Eggwash	
	Wholemeal breadcrumbs	
10 ml	Polyunsatured oil (for cooking)	2 tspns

METHOD

1 Thoroughly mix all the ingredients together.
2 Divide the mixture into portions, then shape into fish cakes using flour to avoid sticking.
3 Pass through the eggwash and coat with breadcrumbs.
4 Neatly reshape.
5 Place the fish cakes on a grill tray and lightly brush over the surfaces with oil.
6 Place under a grill until cooked and lightly coloured. Serve with a suitable sauce, for example low-fat tomato or tartare sauce.

KEBABS

Kofta kebabs

10 portions

Each portion contains:
1735 kJ/411 kcal, 3 g fibre,
13 g fat.

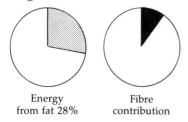

Energy
from fat 28%

Fibre
contribution

850 g	Lean minced lamb or beef	1 lb 14 oz
300 g	Chopped onions	10½ oz
20 g	Ground cumin	2 dspns
30 g	Paprika	1 oz
To taste	Chilli pepper	To taste
5 g	Ground cinnamon	2 tspns
30 g	Fresh chopped parsley	1 oz
30 g	Fresh chopped coriander leaves	1 oz
To taste	Ground black pepper	To taste
5 g	Salt	1 tspn

Accompaniment

1 kg	Riz pilaff (page 128)	2 lb 3 oz

METHOD

1 Thoroughly mix together all the ingredients.
2 Divide the mixture into portions.
3 Shape the mixture around the skewers to form cylinder shapes. *Note:* For large-scale cooking it may be desirable to shape the mixture into sausage shapes without the skewers being used.
4 Grill the kebabs until cooked.
5 Remove the skewers and serve the kebabs accompanied with the riz pilaff and a suitable sauce, for example tomato, curry or Indian-style curry sauce.

Minted lamb kebabs

10 × 100 g portions
Each portion contains:
1846 kJ/438 kcal, 4 g fibre,
14 g fat.

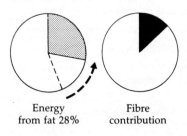

Energy   Fibre
from fat 28%   contribution

1 kg	Diced lean lamb	2 lb 3 oz
500 g	Onion pieces	1 lb 2 oz
40	Mint leaves	40
To taste	Ground black pepper	To taste

Marinade

120 g	Chopped onion	4 oz
2½ g	Crushed garlic	2 cloves
2½ g	Grated lime zest (or lemon)	1 tspn
30 ml	Lime juice (or lemon)	1 fl. oz
30 g	Chopped mint	1 oz
300 ml	Plain natural yogurt	½ pt

Accompaniment

1 kg	Braised millet or riz pilaff	2 lb 3 oz

METHOD

1 Mix together all the ingredients for the marinade in a stainless steel or porcelain container. Alternatively, the ingredients may be liquidized.
2 Put the lamb into the marinade, cover and place aside in a refrigerator for 6–12 hours.
3 Remove the lamb from the marinade and neatly arrange on skewers with the pieces of onion and mint leaves.
4 Season with the black pepper then cook on a grill. During cooking, turn the kebabs as required and brush over the surfaces with the marinade.
5 Serve the kebabs accompanied with braised millet or rice (page 128).

A suitable sauce such as a vegetable raita (page 167) or tsatziki (page 166) may also be served.

TANDOORI-STYLE COOKING

This style of cooking refers to the use of a special clay oven where the chicken is cooked over glowing charcoal embers. However, when no tandoor oven is available, a traditional grill and oven make good substitutes.

Cooking 'Tandoori style' is an excellent method of producing low-fat chicken dishes because the skin is always removed from the chicken prior to cooking and the spicy marinade contains very little fat.

Tandoori chicken

10 portions

Each portion contains:
736 kJ/175 kcal, <1 g fibre, 6 g fat.

Energy
from fat 29%

10	Skinned chicken portions	10
To taste	Ground black pepper	To taste
To taste	Paprika	To taste
To taste	Garam masala	To taste
60 ml	Lemon juice	2 fl. oz
	Red food colour	
	Orange food colour	

Marinade

450 ml	Plain natural yogurt	16 fl. oz
15 g	Garam masala	1½ dspns
10—20 g	Small hot peppers— seeded (quantity to taste)	¼–½ oz
5 g	Salt	1 tspn
Pinch	Ground black pepper	Pinch
30 g	Peeled root ginger	1 oz
120 g	Chopped onion	4 oz
5 g	Garlic	2 cloves
	Zest from 1 lemon	

Accompaniments: These are not included in the nutrient calculations

10	Hot nan bread (page 206)	10
½ litre	Cucumber raita (page 167)	18 fl. oz
10 portions	Indian mixed salad (salat, page 177)	10 portions

See overleaf for method

METHOD

1 Cut slits on the chicken pieces then season with the pepper, paprika and Garam masala.
2 Brush over all the surfaces of the chicken with the lemon juice coloured with a little red and orange colouring.
3 Liquidize all the ingredients for the marinade to a smooth paste.
4 Put the chicken in the marinade, cover and place aside in a refrigerator for 8–12 hours.
5 Remove the chicken from the marinade and cook under a hot grill. Turn when half cooked and complete the cooking. *Important:* If parts of the chicken remain uncooked, complete the cooking of the chicken in the oven.
6 Serve the chicken accompanied with the hot nan bread, raita and salad.

Another excellent accompaniment for this dish is a curry sauce which has been flavoured with the marinade. Simply prepare a plain curry sauce (page 81), then add the marinade from the tandoori chicken. Bring to the boil and simmer for 4–5 minutes.

CHAPTER 8

Cold buffet

It is a simple matter for most cooks with a limited knowledge of nutrition to supply salads which are nutritious and low in fat. However, it is not generally realized that the traditional green salad, although colourful and bulky in appearance, does not contain very much fibre as such small amounts of the ingredients are eaten (see pages 175–179). Neither do salads give a feeling of fullness unless accompanied by a high-fat dressing. To avoid this problem, green salads can be mixed with cooked pulses such as kidney beans and chick-peas, pastas, brown rice and cracked wheat. These improve the colour and texture of a salad as well as increasing the fibre level. The recipes below are just a few examples of the many variations possible. The production of salad dressings, dips, fillings and spreads which are low in fat presents the cook with a much more difficult task. This chapter should enable the reader to prepare a wide range of these items.

SAUCES, SALAD DRESSINGS AND DIPS

Cookery hints

- Salad dressings, sauces and dips should be prepared and served as fresh as possible, particularly when they contain chopped fresh vegetables which quickly deteriorate. Remember that these sauces contain no added preservative; therefore store in a refrigerator for short periods only.
- Try using different combinations of spices, herbs and vegetables to produce unique and interesting flavours.
- Remember that seasoning of sauces, salad dressings and dips can be achieved using spices and herbs as an alternative to salt.
- The use of small spoons for the service of sauces and dressings should help to keep portion sizes to a minimum.
- When preparing crudités or vegetable dips, cut the vegetable as thick as possible; this should ensure that more vegetable than dip is eaten.

Salad dressing 1

2 litres/3½ pt

This salad dressing is a substitute for mayonnaise, which is almost all oil in composition and therefore very high in total fat and percentage energy from fat. This salad dressing still has a high percentage energy from fat, but only contains one-quarter of the fat per portion when compared with mayonnaise, and is ideal for use with composite salads and spreads (see pages 171–179). The use of some olive oil in place of the polyunsaturated oil will enhance the flavour but increase the cost.

2 litres contains:
21,203 kJ/5143 kcal, 4 g fibre, 541 g fat.

Energy
from fat 94%

(A)

1 litre	Water	1¾ pt
250 ml	Polyunsaturated oil	9 fl. oz
200 ml	Vinegar	7 fl. oz
30 ml	Lemon juice	1 fl. oz

(B) *Thickening paste*

500 ml	Cold water	18 fl. oz
60 g	Besan flour	2 oz

(C)

30 g	Mayodan powder	1 oz
250 ml	Polyunsaturated oil	9 fl. oz

(D) Additional flavouring

20 g	Mustard paste	2 tspns
60 g	Tahini paste	2 oz
10 ml	Tamari sauce (optional)	2 tspns
5 ml	Worcester sauce (optional)	1 tspn
20 g	Salt	4 tspns
To taste	Ground white pepper	To taste

METHOD

1 Place the (A) ingredients—the water, oil, vinegar and lemon juice—in a saucepan and bring to the boil.
2 Blend together the (B) ingredients—flour and cold water—until a smooth paste; pass through a fine strainer to ensure that no lumps are present.
3 Whisk the thickening paste into the boiling liquid and bring back to the boil.
4 Boil for 1 minute and whisk to achieve a smooth sauce.
5 Allow to cool.
6 Whisk together the mayodan powder and the oil (C ingredients) until thoroughly combined.
7 Quickly whisk the cold sauce into the oil and mayodan powder until a thick dressing is obtained.
8 Add the flavouring ingredients (D) as required and whisk together until combined.

Salad dressing 2

(not containing
mayodan powder)

2 litres/3½ pt

2 litres contain:
23,140 kJ/5600 kcal, 14 g fibre,
549 g fat.

Energy
from fat 88%

(A)		
1 litre	Water	1¾ pt
500 ml	Polyunsaturated oil	18 fl. oz
200 ml	Vinegar	7 fl. oz
30 ml	Lemon juice	1 fl. oz

(B) *Thickening paste*		
500 ml	Cold water	18 fl. oz
200 g	Besan flour	7 oz

(C) Additional flavouring		
20 g	Mustard paste	2 tspns
60 g	Tahini paste	2 oz
10 ml	Tamari sauce (optional)	2 tspns
5 ml	Worcester sauce (optional)	1 tspn
20 g	Salt	4 tspns
To taste	Ground white pepper	To taste

METHOD

Prepare as in salad dressing 1 but omit the mayodan powder and increase
the besan flour to 200 g/7 oz. Also add all the oil to the water when
bringing to the boil. This produces a slightly thicker dressing which is
ideal for use with spreads—see page 171.

Note: During storage and prior to service, the sauce may require the
addition of a little cold water to correct consistency and produce a smooth
appearance. This procedure is not essential when the recipe containing
the mayodan powder is used.

This lower fat salad dressing may be used in place of mayonnaise (see
Table 8.1; overleaf). All the derivative sauces which are made from
mayonnaise can be made with this dressing.

Table 8.1   Comparison of the fat content of traditional mayonnaises with the salad dressing recipes

For comparison	Fat (g/litre)	Energy from fat (%)
Traditional mayonnaise	800	98
Traditional mayonnaise with an equal quantity of plain yogurt	402	90
Salad dressing 1	270	94
Salad dressing 2	274	88
Propriety brand reduced-calorie mayonnaise	260	85

Herb sauce

1 litre/1 ¾ pt
1 litre contains:
9787 kJ/2367 kcal, 14 g fibre, 243 g fat.

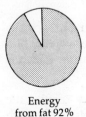

Energy
from fat 92%

Vegetable purée

10 g	Fresh chervil	1 tblspn
120 g	Spinach (fresh or frozen)	4 oz
1 bunch	Fresh watercress	1 bunch
2½ g	Fresh tarragon	½ dspn

Garnish

30 g	Fresh chopped chives	1 oz
5 g	Fresh chopped parsley	1 dspn
To taste	Ground black pepper	To taste
900 ml	Salad dressing 1	1 pt 12 fl. oz

METHOD

1 Prepare the vegetable purée: lightly blanch the vegetables in boiling water; refresh in cold water then squeeze dry.
2 Liquidize the vegetable purée with some of the salad dressing.
3 Mix the vegetable purée with the remaining salad dressing, add the garnish and season with the pepper.

An alternative method of preparing this sauce is to finely chop a selection of fresh herbs and mix through the salad dressing.

Garlic dressing

1 litre/1¾ pt

1 litre contains:
10,601 kJ/2571 kcal, 2 g fibre,
122 g fat.

30 g	Crushed garlic (or simply add to taste)	1 oz
1 litre	Salad dressing 1	1¾ pt
To taste	Ground white pepper	To taste

Energy
from fat 94%

METHOD

Thoroughly mix together all the ingredients.

Thousand island dressing

1 litre/1¾ pt

1 litre contains:
9648 kJ/2339 kcal, 3 g fibre,
243 g fat.

30 g	Finely chopped shallots	1 oz
60 g	Finely chopped red peppers	2 oz
60 g	Finely chopped green peppers	2 oz
30 g	Chopped capers	1 oz
To taste	Fresh chopped parsley	To taste
To taste	Fresh chopped tarragon	To taste
To taste	Tabasco sauce	To taste
900 ml	Salad dressing 1	1 pt 12 fl. oz

Energy
from fat 93%

METHOD

Thoroughly mix together all the ingredients.

Pineapple and curry dressing

1 litre/1¾ pt
1 litre contains:
10,327 kJ/2500 kcal, 5 g fibre, 245 g fat.

Energy
from fat 88%

60 g	Chopped onions	2 oz
5 g	Garlic	2 cloves
5 g	Chilli peppers (without the seeds)	1 small pepper
30 g	Peeled root ginger	1 oz
30 g	Tomato purée	1 oz
To taste	Curry paste	To taste
900 ml	Salad dressing 1	1 pt 12 fl. oz
Garnish		
200 g	Small diced pineapple	7 oz
5 g	Fresh chopped coriander	1 dspn

METHOD

1   Liquidize the onion, garlic, chillis, ginger, tomato purée and curry paste with some of the salad dressing.
2   Mix the puréed vegetables with the remaining salad dressing then add the garnish.

Tsatziki dressing/dip (Greek)

1 litre/1¾ pt
1 litre contains:
1029 kJ/260 kcal, 2 g fibre, 4 g fat.

Energy
from fat 14%

600 g (1 large)	Cucumber—including skin	1 lb 5 oz
10 g	Garlic	4 cloves
2½ g	Lemon zest	1 tspn
10 ml	Lemon juice	2 tspns
To taste	Ground black pepper	To taste
350 ml	Low-fat natural yogurt	12 fl. oz

METHOD

1 Liquidize the ingredients with approximately one-quarter of the yogurt.
2 Add the remaining yogurt and blend together.

Cucumber raita (Indian) (1 litre/1¾ pt)

Prepare the same as Tsatziki above, adding to taste: chilli pepper, fresh chopped coriander and fresh chopped mint.

Vegetable raita

1 litre/1¾ pt

1 litre contains:
2096 kJ/496 kcal, 26 g fibre,
5 g fat.

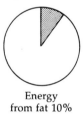

Energy
from fat 10%

120 g	Cucumber (with skins) (¼ large cucumber)	4 oz
60 g	Diced onions	2 oz
10 g	Garlic	4 cloves
120 g	Diced carrots	4 oz
90 g	Diced celery	3 oz
120 g	Cooked haricot beans	4 oz
2½ g	Lemon zest	1 tspn
30 ml	Lemon juice (½ lemon approx.)	1 fl. oz
400 ml	Low-fat natural yogurt	14 fl. oz
To taste	Ground black pepper	To taste
Garnish		
To taste	Chopped chives	To taste
To taste	Chopped mint leaves	To taste
To taste	Chopped coriander	To taste

METHOD

1 Liquidize the ingredients with approximately one-quarter of the yogurt.
2 Add the remaining yogurt and garnish and blend together.

Hummus

1 litre/1¾ pt

Each litre contains:
4899 kJ/1175 kcal, 37 g fibre, 47 g fat.

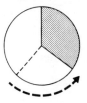

Energy
from fat 35%

500 g	Cooked chick-peas	1 lb 2 oz
20 g	Garlic	8 cloves
60 ml	Lemon juice	2 fl. oz
60 g	Tahini (sesame seed and oil paste—available commercially)	2 oz (2 tblspns)
	Bean cooking liquor (to adjust consistency)	

METHOD

1 Liquidize the chick-peas, garlic, lemon juice and tahini with some of the cooking liquor. Use enough cooking liquor to produce a thick dressing.
2 Adjust consistency to taste.

Chick-pea and yogurt dressing

1 litre/1¾ pt

1 litre contains:
3967 kJ/933 kcal, 41 g fibre, 17 g fat.

Energy
from fat 16%

500 g	Cooked chick-peas	1 lb 2 oz
30 ml	Lemon juice	1 fl. oz
5 g	Lemon zest	1 tspn
200 ml	Plain natural yogurt	7 oz
To taste	Ground black pepper	To taste
	Bean cooking liquor (to adjust consistency)	
Garnish		
120 g	Finely diced green peppers	4 oz
120 g	Finely diced celery	4 oz

METHOD

1 Liquidize the chick-peas, lemon juice, lemon zest and yogurt with some of the bean cooking liquor. Use enough cooking liquor to produce a thick dressing.
2 Mix through the garnish and adjust consistency to taste with the bean cooking liquor.

Guacamole ★★

1 litre/1¾ pt

1 litre contains:
5324 kJ/1284 kcal, 15 g fibre, 113 g fat.

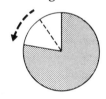

Energy
from fat 78%

500 g	Ripe avocado flesh (3 medium avocados)	1 lb 2 oz
5 g	Garlic	2 cloves
200 g	Tomatoes (chopped with skins and seeds)	7 oz
30 ml	Lemon juice	1 fl. oz
200 ml	Low-fat natural yogurt	7 oz
Garnish		
60 g	Finely chopped green peppers	2 oz
60 g	Finely chopped red peppers	2 oz
10 g	Finely chopped chives	1 dspn
10 g	Finely chopped spring onions	1 dspn
To taste	Chilli power	To taste
To taste	Ground black pepper	To taste

METHOD

1 Liquidize the avocado flesh, garlic, tomatoes and lemon juice with half of the yogurt.
2 Remove from the liquidizer, blend through the remaining yogurt.
3 Mix through the garnish, season with the black pepper.

Avocado dressing/purée/ filling★★

1 litre/1¾ pt

1 litre contains:
5563 kJ/1340 kcal, 17 g fibre,
111 g fat.

500 g	Ripe avocado flesh (3 medium avocados)	1 lb 2 oz
120 g	Cooked peas	4 oz
30 g	Onion	1 oz
5 g	Parsley	2 tspns
30 ml	Lemon juice	1 fl. oz
350 ml	Fromage frais	12 fl. oz
To taste	Ground white pepper	To taste
To taste	Cayenne pepper	To taste

Energy
from fat 74%

METHOD

Liquidize the ingredients to a thick purée. Adjust consistency to taste.

Fruit dressings

These can be made by liquidizing fruits to a thin purée, for example pineapples, mangoes, peaches, etc. The fruit purées may require the addition of lemon juice, depending on the type of fruit used. Yogurt may also be used as a base for fruit dressings.

Orange dressing

1 litre/1¾ pt

1 litre contains:
2059 kJ/494 kcal, fibre negligible, 8 g fat.

200 ml	Orange juice	7 fl. oz
30 ml	Lemon juice	1 fl. oz
30 g	Grated lemon zest	1 oz
20 g	Blanched strips orange zest	1 tblspn
800 ml	Low-fat natural yogurt (thick)	1 pt 8 fl. oz

Energy
from fat 14%

METHOD

1   Boil down the orange and lemon juice until thick and concentrated, allow to cool.
2   Mix all the ingredients together.

Acidulated dressings

These may be made in the same manner as traditional vinaigrette (oil and vinegar dressing) but with a much lower proportion of oil. Use two parts vinegar or lemon juice, etc. to one part oil. Various flavourings may also be added, for example mustard, spices, fresh chopped herbs or vegetables, etc. When it is desirable to have a lower acid content, replace some of the lemon juice or vinegar with low-fat yogurt.

FILLINGS AND SPREADS

These may be made by blending together different commodities with salad dressing 2. Examples are low-fat cheeses, cooked lean meat purées, fish and shellfish purées, and vegetable mixtures. They can be used in sandwiches, filled rolls or piped on to canapés and do not require additional butter or margarine.

Use 30 g/1 oz of spread per round of sandwiches and garnish with salad.

Cheese spread sandwiches

2 slices	Wholemeal bread	2 slices
30 g	Cheese spread	1 oz

1 portion/1 round
1 sandwich contains:
889 kJ/212 kcal, 5 g fibre,
6 g fat.

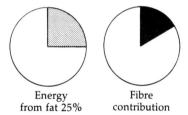

Energy
from fat 25%

Fibre
contribution

Cheese spread

1 kg/2 lb 3 oz
30 portions
1 kg contains:
10,191 kJ/2426 kcal, 1 g fibre,
132 g fat.

850 g	Grated low-fat cheese	1 lb 14 oz
150 ml	Salad dressing 2	¼ pt
To taste	Ground white pepper	To taste
To taste	Cayenne pepper (optional)	To taste

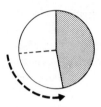

Energy
from fat 47%

METHOD

Mix all the ingredients together to produce a thick spreading paste.

Variations

Cheese and vegetable spread

1 kg contains:
8599 kJ/2047 kcal, 5 g fibre,
103 g fat.

60 g	Chopped onions	2 oz
60 g	Grated carrots	2 oz
30 g	Finely diced celery	1 oz
60 g	Finely diced red and green peppers	2 oz
To taste	Chopped parsley	To taste

Energy
from fat 44%

METHOD

Reduce the cheese content in the recipe for cheese spread to 650 g/1 lb 7 oz and add the above ingredients.

Cottage cheese spreads

A variety of tasty spreads may be made with a very low fat content by adding different ingredients to cottage cheese, for example, freshly chopped herbs, chives, cut fresh vegetables (peppers, cucumber and spring onions), blanched vegetables (carrots, celery, onions, courgettes and cauliflower), grated root ginger and spices (commercial spice pastes). Cooked lean meats, fish and shellfish may also be added to cottage cheese to produce interesting spreads.

Meat and poultry spreads

Ham, beef, chicken, turkey, etc.

1 kg contains:
7198 kJ/1714 kcal, 1 g fibre, 56 g fat.

*Energy
from fat 29%

*The nutrient analysis used an average figure for meat and poultry. Individual calculations are given in Appendix 2.

850 g	Appropriate cooked meat or poultry	1 lb 14 oz
150 ml	Salad dressing 2	¼ pt
To taste	Ground pepper	To taste

METHOD

Mince the cooked meat and bind with the salad dressing. Alternatively, a food processor may be used.

Fish and shellfish spreads

White fish (cod, whiting or haddock), prawn, lobster or crab, etc.

1 kg contains:
5795 kJ/1380 kcal, 1 g fibre, 25 g fat.

900 g	*Cooked fish or shellfish	2 lb
120 ml	Salad dressing 2	4 fl. oz
To taste	Ground white pepper	To taste
To taste	Cayenne pepper	To taste

*Cooked fish and shellfish is often mixed together to produce fish and shellfish spreads. Finely chopped onion, diced cucumber and chopped chives, etc. are also appropriate to these spreads.

Energy
from fat 10%

Vegetable spread

1 kg/2 lb 3 oz

1 kg contains:
2598 kJ/617 kcal, 24 g fibre, 18 g fat.

Energy
from fat 26%

90 g	Chopped onion	3 oz
90 g	Chopped celery	3 oz
90 g	Grated carrots	3 oz
120 g	Small dices of red and green peppers	4 oz
150 g	Small dices of cucumber	5 oz
150 g	Sweetcorn kernels	5 oz
150 g	Bean sprouts	5 oz
30 g	Small dices of gherkin	1 oz
30 g	Sunflower seeds	1 oz
200 ml	Chick-pea dressing	7 fl. oz
To taste	Black pepper	To taste

METHOD

Mix together all the ingredients using enough chick-pea dressing to bind the mixture and produce a good spreading consistency.

SALADS

Arabian salad

10 portions—large

Each portion contains:
170 kJ/40 kcal, 2 g fibre,
2 g fat.

200 g	Picked and washed watercress (2 bunches approximately)	7 oz
750 g	Tomato pieces	1 lb 10 oz
600 g	Diced cucumber (1 large cucumber)	1 lb 5 oz
300 g	Chopped onions	10½ oz
Dressing		
30 ml	Lemon juice	1 fl. oz
15 ml	Polyunsaturated oil	½ fl. oz
To taste	Tabasco sauce	To taste

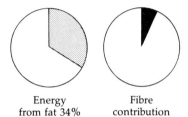

Energy from fat 34%   Fibre contribution

METHOD

1   Prepare the dressing by whisking together the ingredients.
2   Mix all the ingredients with the dressing and serve.

Baba ganaouge

10 portions
A spicy aubergine salad from
Armenia.

Each portion contains:
115 kJ/27 kcal, 2 g fibre,
1 g fat.

500 g	Aubergine (1 large aubergine)	1 lb 2 oz
250 g	Strips of green pepper	9 oz
250 g	Thinly sliced tomatoes	9 oz
140 g	Thinly sliced onions	5 oz
Dressing		
10 g	Cummin powder	3 tspns
10 g	Crushed garlic	4 cloves
5 g	Salt	1 tspn
To taste	Cayenne pepper	To taste
30 ml	Lemon juice	1 fl. oz
10 g	Fresh chopped parsley	1 dspn
10 ml	Olive oil	2 tspns

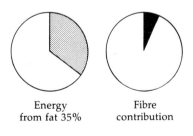

Energy from fat 35%   Fibre contribution

see overleaf for method

METHOD

1 Pierce the aubergine once with a sharp knife and place whole and unpeeled on a shelf in a hot oven until it is soft when pressed.
2 Remove from the oven, allow to cool, then peel off the skin.
3 Chop the flesh then add the green peppers, onions and tomatoes.
4 Prepare the dressing by thoroughly mixing together all the ingredients.
5 Bind the aubergine mixture with the dressing, serve chilled garnished with a little extra chopped parsley.

Bean salad

10 portions—large

Each portion contains:
363 kJ/85 kcal, 6 g fibre, 0.5 g fat.

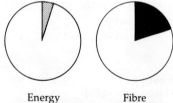

| Energy | Fibre |
| from fat 5% | contribution |

800 g	Cooked beans (370 g/13 oz raw beans)	1 lb 12 oz
120 g	Sliced onions	4 oz
5 g	Crushed garlic	2 cloves
120 g	Diced tomato flesh	4 oz
10 g	Chopped parsley	1 dspn
5 g	Chopped basil	2 tspns
2½ g	Allspice	1 tspn
To taste	Ground black pepper	To taste
120 ml	Low-fat plain yogurt (to bind)	4 fl. oz

METHOD

Mix together the beans, onions, garlic, tomato and fresh herbs. Season with the spice and pepper then bind with the yogurt.

Rice and vegetable salad with orange dressing

10 portions—large

Each portion contains:
541 kJ/127 kcal, 3 g fibre,
1 g fat.

750 g	Lightly cooked brown rice (300 g/10 oz raw rice)	1 lb 10 oz
120 g	Diced cucumber	4 oz
120 g	Diced red and green peppers	4 oz
120 g	Diced carrots	4 oz
120 g	Diced celery	4 oz
60 g	Raw cauliflower florets	2 oz
60 g	Sweetcorn kernels	2 oz
60 g	Sliced red radishes	2 oz
100 g	Orange segments	2 oranges
200 ml	Orange dressing—to bind (page 170)	7 fl. oz

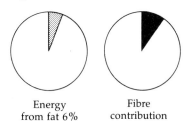

Energy from fat 6% • Fibre contribution

METHOD

Mix together all the ingredients and bind with the orange dressing.

Salat (Indian)

10 portions—large

Each portion contains:
68 kJ/16 kcal, 1.5 g fibre,
<1 g fat.

400 g	Shredded lettuce	14 oz
150 g	Strips of onion	5 oz
200 g	Sliced cucumber	7 oz
400 g	Sliced tomatoes	14 oz
To taste	Fresh coriander leaves	To taste

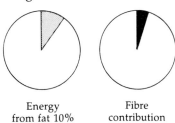

Energy from fat 10% • Fibre contribution

METHOD

Mix together all the ingredients and neatly arrange in a salad bowl.
Decorate with the coriander leaves and serve.

Egyptian bread salad

10 portions—large

Each portion contains:
283 kJ/67 kcal, 3 g fibre,
2 g fat.

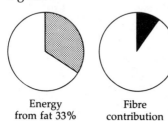

Energy   Fibre
from fat 33%   contribution

400 g	Shredded lettuce (firm type)	14 oz
150 g	Diced cucumber	5 oz
300 g	Sliced tomatoes	10½ oz
120 g	Diced green peppers	4 oz
30 g	Chopped spring onions	1 oz
10 g	Chopped coriander leaves	1 dspn
10 g	Chopped parsley	1 dspn
10 g	Chopped mint leaves	1 dspn
2½ g	Crushed garlic	1 clove
200 g	Toasted wholemeal bread (thin rectangles)	7 oz

Dressing

60 ml	Lemon juice	2 fl. oz
10 ml	Polyunsaturated oil	2 tspns
	Ground black pepper	To taste

METHOD

Mix together all the ingredients *except* the bread and bind with the dressing. Add the toasted bread and combine with the salad when serving.

Moroccan carrot salad

10 portions—large

Each portion contains:
100 kJ/24 kcal, 3 g fibre.

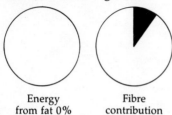

Energy   Fibre
from fat 0%   contribution

1 kg	Raw carrot strips (or grated)	2 lb 3 oz
2½ g	Ground cinnamon	1 tspn
90 ml	Lemon juice	3 fl. oz
30 ml	Rose water	1 fl. oz

METHOD

Mix the cinnamon and carrots, bind with the lemon juice and rose water.

Pasta salad

10 portions—large

Each portion contains:
422 kJ/100 kcal, 4 g fibre,
1 g fat.

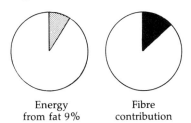

Energy
from fat 9%

Fibre
contribution

750 g	Lightly cooked wholemeal pasta— macaroni, shells, etc. (use 300 g/10 oz raw pasta)	1 lb 10 oz
60 g	Diced onions	2 oz
60 g	Diced cucumber	2 oz
120 g	Diced carrots	4 oz
120 g	Diced red and green peppers	4 oz
120 g	Diced celery	4 oz
To taste	Ground black pepper	To taste
150 ml	Tsatziki dressing— to bind (page 166)	5 fl. oz

METHOD

Mix together all the ingredients and bind with the dressing.

Turkish shepherd's salad

10 portions

Each portion contains:
175 kJ/41 kcal, 3 g fibre,
<1 g fat.

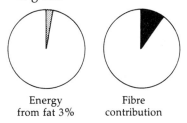

Energy
from fat 3%

Fibre
contribution

500 g	Diced crisp dessert apples	1 lb 2 oz
250 g	Diced cucumber (with skins) (½ cucumber approximately)	9 oz
250 g	Diced tomatoes	9 oz
250 g	Diced celery	9 oz
250 g	Diced green peppers	9 oz
250 g	Diced onions	9 oz
10 g	Chopped parsley	1 dspn
Dressing		
50 ml	Lemon juice	2 lemons

METHOD

Mix the apple and lemon juice together to coat the surfaces thoroughly and prevent the apple going brown. Mix through the remaining ingredients and serve garnished with chopped parsley.

CHAPTER 9

Pastries, sweets and puddings

The sweet section of a menu is often less appealing because of the richness of the desserts offered, especially when heavily decorated with cream. Lighter desserts and more emphasis on fruit and fruit fillings will both be healthier and increase the popularity of desserts.

The recipes in this section have a reduced total fat content and incorporate polyunsaturates instead of saturates whenever possible. Savoury dishes made with doughs and pastry are also included in this chapter.

Note: The pastry recipes do not all have 35% energy from fat when calculated on their own but are an improvement on standard pastry recipes. When fillings are included the percentage energy from fat falls below 35%, with the exception of the quiche containing cheese.

SHORT PASTRY AND SHORT PASTE PRODUCTS

The quantity of fat in short pastry is important because it contributes a crisp tender texture. In the following recipes the fat content is 37.5% the weight of the flour and produces a good all-round pastry, especially when baking blind. However, when a short pastry is cooked together with a filling, as for example with Cornish pasties, quiches, fruit tarts, etc., the fat may be reduced to 25% the weight of the flour, with a corresponding increase in the recipe water.

Cookery hints

To reduce the likelihood of producing tough pastry with a hard eating texture:

- Use a very weak flour, for example flour milled from English wheats.
- Avoid over-handling.
- Allow the pastry to rest after handling.

Sweet short pastry 1

(for baking blind)

Each recipe contains*:
12,222 kJ/2910 kcal,
14 g fibre, 154 g fat.

400 g	Plain flour	14 oz
5 g	Baking powder	1 tspn
150 g	Polyunsaturated fat	5 oz
40 g	Castor sugar	2 tbspns
100 ml	Water	3½ fl. oz

Energy
from fat 46%

*Nutrient analysis is based on the use
of sunflower oil.

METHOD

1 Sieve together the flour and baking powder.
2 Add the fat and rub into a sandy texture.
3 Dissolve the sugar in the water, add and mix together to combine the ingredients.

Savoury short pastry 1

Prepare as sweet short pastry 1 but replace the sugar with 2 g/pinch salt.

Sweet short pastry 2

(cooked with a filling, e.g., fruit tarts and pies and filled pastries)

Each recipe contains*:
9678 kJ/2295 kcal, 14 g fibre, 86 g fat.

400 g	Plain flour	14 oz
5 g	Baking powder	1 tspn
100 g	Polyunsaturated fat	3½ oz
40 g	Castor sugar	2 tbspns
130 ml	Water	4½ fl. oz

Energy
from fat 33%

*Nutrient analysis is based on the use of polyunsaturated margarine.

METHOD

1   Sieve together the flour and baking powder.
2   Cream together the fat and an equal quantity of the flour, that is 100 g, until very light.
3   Lightly rub in the remaining flour.
4   Dissolve the sugar in the water, add and mix together to combine the ingredients.

Savoury short pastry 2

(cooked with a filling, e.g., Cornish pasties and quiches)

Prepare as sweet short pastry 2 but replace the sugar with 2 g/pinch salt.

Wholemeal short pastry

400 g	Wholemeal flour	14 oz
5 g	Baking powder	1 tspn
120 g	Polyunsaturated fat	4 oz
Pinch (2 g)	Salt	Pinch
140 ml	Water	5 oz

Each recipe contains:
9038 kJ/2156 kcal, 38 g fibre, 105 g fat.

Energy
from fat 33%

METHOD

1 Thoroughly mix together the flour and baking powder.
2 Add the fat and rub into a sandy texture.
3 Dissolve the salt in the water, add and mix together to combine the ingredients.

Note: To estimate the correct quantity of water for large-scale recipes, add two-thirds of the water and mix through, then correct the consistency with the remaining one-third, added in stages.

Cookery hints

- Wholemeal flour contains small particles of bran, the outer husk of wheat, which is rich in fibre, therefore it should not be sifted before use.
- The bran in wholemeal flour absorbs water fairly slowly. If any difficulty is experienced when adding water to a wholemeal pastry, allow the mixture to rest for a few minutes between each addition. Too little water will result in a dry, crumbly crust.
- Finely ground wholemeal flours will take up more water than courser ones.

FRUIT FLANS, PIES AND TARTS

Fruit flans

10 portions
(225 mm/9 inch diameter flan ring)
Each portion contains:
1036 kJ/246 kcal, 2 g fibre, 9 g fat.

350 g	Sweet short pastry 1 (page 181)	12 oz
500 g	Prepared fresh or *tinned fruit: apricots, pears, pineapple, etc.	1 lb 2 oz
500 ml	Pastry cream (see below)	18 fl. oz
75 ml	Fruit glaze (drained fruit juice thickened with arrowroot)	2½ fl. oz

*Unsweetened tinned fruit—lightly sweetened to taste.

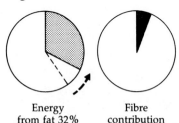

Energy from fat 32%   Fibre contribution

METHOD

1 Line the flan ring with the pastry and bake 'blind', allow to cool.
2 Prepare the fruit ready for filling the flan; leave small fruits whole and cut large fruits into slices or suitable pieces.
3 Spread the pastry cream over the base of the flan then neatly arrange the fruit on top.
4 Prepare the glaze then thinly mask the hot glaze over the top of the fruit.

Pastry cream

Yield 500 ml/18 fl. oz approximately

40 ml	Egg yolks	2 egg yolks
60 g	Castor sugar	2 oz
40 g	Plain flour	1½ oz
400 ml	Skimmed milk	14 fl. oz
3–4 drops	Drops of vanilla essence	3–4 drops

METHOD

1 Cream together the egg yolks and sugar, add the essence.
2 Mix in the flour.

3 Boil the milk then whisk onto the sugar, egg yolks and flour—ensure the ingredients are well mixed.
4 Bring the mixture to the boil, stirring continuously, then simmer for 1 minute. Avoid burning the mixture.
5 Cool quickly and store under cover until required for use.

Lemon meringue flan

10 portions
(225 mm/9 inch diameter flan ring)

Each portion contains:
1368 kJ/326 kcal, 1 g fibre, 8 g fat.

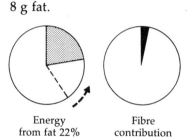

Energy from fat 22%   Fibre contribution

350 g	Sweet short pastry 1 (page 181)	12 oz
Lemon curd		
425 ml	Water	15 fl. oz
120 g	Sugar	4 oz
75 g	Besan flour	2½ oz
4	Lemons	4
15 g	Polyunsaturated margarine	½ oz
Meringue		
120 ml	Egg whites	4 fl. oz
200 g	Castor sugar	7 oz

METHOD

1 Line the flan ring with the pastry and bake blind.

Preparation of filling
2 Blend together the besan flour with 125 g/4 fl. oz of the water to form a smooth paste.
3 Bring to the boil the remaining water and add the sugar.
4 Pour in the thickening paste and stir to obtain a very thick smooth paste. Cook for 1 minute, avoiding burning, then remove from the heat.
5 Grate the lemon zest and extract the juice.
6 Add the lemon juice and zest, mix together.
7 Blend in the margarine.
8 Place the filling into the flan case.
9 Prepare a meringue with the egg whites and castor sugar. Pipe neatly across the top of the flan.
10 Place the flan back in the oven, bake until a good colour.

Fruit pies

10 portions

Each portion contains:
812 kJ/191 kcal, 3 g fibre,
4 g fat.

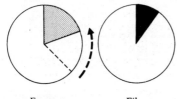

Energy   Fibre
from fat 20%   contribution

350 g	Sweet short pastry 2 (page 182)	12 oz
800 g	Prepared fruit, e.g., apples, gooseberries, rhubarb	1 lb 12 oz
*120 g	Castor sugar	*4 oz
*60 ml	Water	*2 fl oz
	Egg wash	

*The exact quantity of sugar and water will depend on the type of fruit used. Remember to keep the sugar low.

METHOD

1 Wash and prepare the fruit, mix together with the water and sugar.
2 Place the fruit mixture into a suitable pie dish; the fruit mixture should fill the pie dish.
3 Cover with the pastry.
4 Notch or mark the edges of the pastry and cut a small hole in the centre. Brush over with the egg wash.
5 Place the pie on a baking tray and bake at 200°C/390°F for 40 minutes approximately.
6 Accompany with 500 ml/18 fl. oz skimmed milk custard sauce (page 215).

Fruit turnovers

Same recipe as for fruit pies above, but increase the pastry to 450 g/1 lb.

METHOD

1 Wash and prepare the fruit.
2 Lightly cook the fruit with the water and sugar. Allow to cool.
3 Roll out the pastry and cut into squares (110 mm/4 inch sides).
4 Lightly dampen the edges of the pastry with a little water.
5 Place the fruit filling onto the centre of each square, neatly fold over to make a triangle. Press down to achieve a good seal.
6 Bake at 200°C/390°F for 30 minutes approximately.
7 Accompany with 500 ml/18 fl. oz skimmed milk custard sauce (page 215).

Fruit tarts

(10 portions)

Each portions contains:
1053 kJ/249 kcal, 3 g fibre,
6 g fat.

450 g	Sweet short pastry 2 (page 182)	1 lb
800 g	Prepared fruit, e.g., apples, gooseberries, rhubarb	1 lb 12 oz
*120 g	Castor sugar	*4 oz
*60 ml	Water	*2 fl oz
	Egg wash	

*The exact quantity of sugar and water will depend on the type of fruit used. Remember to keep the sugar low.

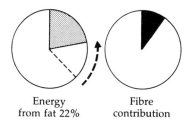

Energy from fat 22%   Fibre contribution

METHOD

1   Wash and prepare the fruit, lightly stew in a saucepan with the sugar and water. Allow to cool.
2   Line a flan ring or lightly greased tart plate with two-thirds of the pastry; ensure that the edge of the ring or plate is overlapped with pastry.
3   Place the fruit into the pastry case and lightly dampen around the edge of the pastry with water.
4   Roll out the remaining pastry and cover over the top.
5   Press down to seal the edges and neatly notch or mark.
6   Cut a neat hole in the top of the tart then brush over with egg wash.
7   Bake at 200°C/390°F for 30 minutes approximately.
8   Allow to cool slightly then carefully detach from the ring or plate.

Baked apple dumplings

10 portions

Each portion contains:
1491 kJ/355 kcal, 7 g fibre,
9 g fat.

700 g	Sweet short pastry 2 (page 182)	1 lb 9 oz
10	Small cooking apples (or halves of large apples)	10
75 g	Castor sugar	2½ oz
Pinch	Ground cloves	Pinch
	Egg wash	

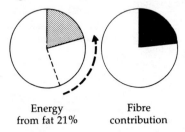

Energy
from fat 21%

Fibre
contribution

METHOD

1 Peel, core and wash the apples then place into lemon water.
2 Roll out the pastry then cut into shapes large enough to cover each apple.
3 Envelope each apple in pastry. Decorate the tops with leaves if desired.
4 Place onto a lightly greased baking tray and brush over with egg wash.
5 Bake at 200°C/390°F for 30 minutes approximately.
6 Accompany with 500 ml/18 fl. oz skimmed milk custard sauce (page 215).

Note: Dessert apples or pears can be substituted for cooking apples. They result in a more compact portion and often need less sugar.

SAVOURY DISHES USING SHORT PASTRY

Quiches

Quiches are popular and can be made with many different vegetable fillings. The following recipe may be used as a guide for producing a range of vegetable-type quiches. Wholemeal pastry can be substituted for short pastry if preferred.

Cookery hints

- Use thin black metal baking trays for crisper pastry.
- When using vegetables which contain a large proportion of water, for example tomatoes, use a little less skimmed milk (reduce milk in recipe to 250 m/9 fl. oz).
- Do not shallow fry or sweat vegetables which are to fill a quiche; this only adds unnecessary fat. If a crisp vegetable texture is not desired, simply lightly blanch or steam the vegetables.

*Vegetable quiche**★★*

8 portions
(225 mm/9 inch diameter flan ring)

Each portion contains:
1013 kJ/241 kcal, 1.5 g fibre, 12 g fat.

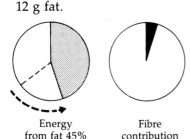

Energy
from fat 45%

Fibre
contribution

350 g	Savoury short pastry 2 (page 182)	12 oz
Filling		
120 g	Chopped onion	4 oz
2½ g	Garlic	2 cloves
90 g	Diced green pepper	3 oz
60 g	Diced blanched carrots	2 oz
60 g	Cooked brown rice	2 oz
60 g	Low-fat cheddar-type cheese	2 oz
Savoury custard		
300 ml	Skimmed milk	10½ fl. oz
100 ml	Eggs	2 (size 3)

METHOD

1 Line the flan ring with the pastry and fill with the vegetables.
2 Mix together the eggs and milk and pour into the flan.
3 Sprinkle over the cheese and bake at 190°C/380°F.

Various quiches
Quiches may be made to incorporate a whole range of vegetables, for example diced red and green peppers, sweetcorn kernels, cooked brown rice, diced onions, diced blanched carrots, sliced tomatoes.

Pasties

Cornish mince and vegetable pasties

10 portions

Each portion contains:
1484 kJ/356 kcal, 7 g fibre, 14 g fat.

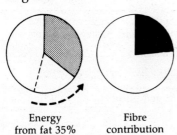

Energy
from fat 35%

Fibre
contribution

800 g	Wholemeal short pastry (page 183)	1 lb 12 oz
200 g	Lean minced beef	7 oz
120 g	Diced onions	4 oz
300 g	Peeled diced potatoes	10½ oz
60 g	Blanched diced celery	2 oz
90 g	Diced blanched carrots	3 oz
90 g	Cooked sweetcorn kernels	3 oz
	Eggwash	
5 g	Salt	1 tspn

METHOD

1 Roll out the pastry to a thickness of 4 mm/⅜ inch, cut into rounds with a 160 mm/6 inch diameter plain round cutter.
2 Prepare the filling by thoroughly mixing together the beef, onions, potatoes, celery, carrots and sweetcorn.
3 Place the filling onto the pastry rounds (75 g/2½ oz each), lightly dampen around the edges with water.
4 Fold the pastry in half, bringing the edges upwards over the filling, then seal together.
5 Notch the tops neatly, then place onto a lightly greased baking tray.
6 Brush over the tops with egg wash and bake at 200°C/390°F for 40 minutes approximately.
7 Accompany with a suitable sauce, for example 500 ml/18 fl. oz tomato sauce (page 81).

CHOUX PASTE★★

This recipe substitutes polyunsaturated oil for margarine, but does not appreciably alter the total fat content.

Cookery hints

- Before adding any egg, allow the mixture to cool to a temperature which will not cause any of the egg to cook.
- Adding egg to achieve the correct consistency of paste is important. The paste should be soft with the ability to hold it's shape (just and no more) when piped.
- Ensure the paste is thoroughly cooked and crisp before removing from the oven. Under-cooked pastry will collapse when removed from the oven.

Basic recipe

Each recipe contains:
5617 kJ/1337 kcal, 5 g fibre, 79 g fat.

Energy
from fat 52%

250 ml	Water	9 fl. oz
50 ml	Polyunsaturated oil	2 tblspns
150 g	Plain flour	5 oz
250 ml	Eggs	5 (size 3)

METHOD

1 Sieve the flour to remove any lumps.
2 Place the water and oil in a saucepan and bring to the boil.
3 Add all the flour at one time and mix over a low heat for 1 minute. The mixture should leave the sides of the pot cleanly.
4 Allow to cool slightly (see cookery hints above).
5 Add the eggs a little at a time while beating the paste. The eggs should be added until a soft dropping consistency is achieved (see cookery hints above).
6 When piped into appropriate shapes onto a greased tray, bake at 220°C/420°F until cooked.

PUDDING PASTE AND PUDDING PASTE PRODUCTS

An excellent alternative for suet pastry.

Basic recipe

Each recipe contains:
9804 kJ/2334 kcal, 14 g fibre,
104 g fat.

400 g	**Plain flour**	7 oz
20 g	**Baking powder**	¾ oz
100 ml	**Vegetable oil**	3½ fl. oz
140 ml	**Cold water**	5 fl. oz

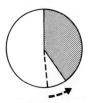

Energy
from fat 40%

METHOD

1 Sieve together the flour and baking powder.
2 Whisk the oil and water together, add to the flour and lightly mix to combine the ingredients.

STEAMED FRUIT PUDDINGS

Steamed apple and sultana pudding

10 portions

Each portion contains:
1321 kJ/313 kcal, 4 g fibre,
10 g fat.

650 g	Pudding paste (basic recipe)	1 lb 7 oz
700 g	Peeled sliced apples	1 lb 9 oz
120 g	Sultanas	4 oz
60 g	Castor sugar	2 oz
30 ml	Water	1 fl. oz

Energy
from fat 29%

Fibre
contribution

METHOD

1 Lightly grease a suitable pudding basin or cooking dish, line with three-quarters of the paste.
2 Mix the ingredients for the filling together, then place into the lined basin, leaving a 10 mm/½ inch gap from the filling to the top of the lined basin.
3 Dampen around the edges of the paste with water, then roll out the remaining paste and cover the top. Seal around the edges, neatly trim if required.
4 Cover with lightly greased greaseproof paper and a pudding cloth, which is secured with string and with the ends tied together.
5 Steam for 1½–2 hours.
6 Accompany with 500 ml/18 fl. oz skimmed milk custard sauce (page 215).

SAVOURY DISHES USING PUDDING PASTE

Savoury dumplings
10 portions

These offer a suitable alternative for potatoes and rice. The dumplings may be plain, or different vegetables cut into small pieces can be added to the dough prior to shaping. Spices, herbs, nuts and fruit (raisins, sultanas, prunes, etc.) may be added to produce unique combinations of taste, flavour and texture.

METHOD

Prepare the basic recipe for pudding paste, then shape into tight balls. The dumplings may be steamed, poached or added to a stew near the end of the cooking period.

Steak and vegetable pudding
10 portions

Each portion contains:
1482 kJ/352 kcal, 3 g fibre, 14 g fat.

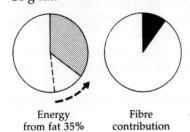

Energy from fat 35%   Fibre contribution

650 g	Pudding paste (page 192)	1 lb 7 oz
800 g	Diced lean stewing steak	1 lb 12 oz
15 ml	Polyunsaturated oil	½ fl. oz
250 g	Diced onions	9 oz
120 g	Diced celery	4 oz
120 g	Diced carrots	4 oz
120 g	Sliced mushrooms	4 oz
Good pinch	Chopped parsley	Good pinch
15 ml	Worcester sauce	½ fl. oz
1 litre	Brown stock	1¾ pt
5 g	Salt	1 tspn

Thickening paste

30 g	Wholemeal flour	1 oz
60 ml	Cold water	2 fl. oz

METHOD

1 Heat the oil in a saucepan then sear the meat. Drain off all excess fat.
2 Add the brown stock, bring to the boil, then slowly simmer. Skim and top up with additional stock as required.
3 When the meat is three-quarters cooked, thicken with the flour and cold water blended together.
4 Add the onion, celery and carrot and lightly cook.
5 Add the mushrooms and parsley then flavour to taste with the Worcester sauce. *Do not completely cook the vegetables.*
6 Cool quickly and remove any surface fat which solidifies on the surface of the sauce.
7 Line a pudding basin with the paste, add the cold mixture, then cover with the paste ready for cooking (see steamed apple pudding, steps 3 and 4, page 193).
8 Steam for 1½–2 hours.

PASTA (white)

Plain and wholemeal pasta recipes suitable for a whole range of pasta dishes: noodles (nouilles), ravioli, cannelloni and lasagne.

Cookery hints

- Mix all the ingredients thoroughly together to produce a smooth, pliable dough.
- Always allow to rest before use or the pasta will shrink when handled.
- Test the pasta regularly during cooking. Fresh pasta cooks very quickly and breaks up when over-cooked.

Plain pasta

10 portions

Each portion contains:
769 kJ/182 kcal, 2 g fibre,
6 g fat.

200 g	*Besan flour	7 oz
200 g	Bread flour	7 oz
50 ml	Polyunsaturated oil	1¾ fl. oz
60 ml	Egg white	2 egg whites
90 ml	Water	3 fl. oz
5 g	Salt	1 tspn

*This may be replaced with bread flour but the
fibre content will be reduced.

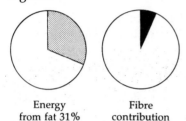

Energy
from fat 31%

Fibre
contribution

Wholemeal pasta

10 portions

Each portion contains:
734 kJ/174 kcal, 4 g fibre,
6 g fat.

400 g	Wholemeal flour	14 oz
50 ml	Polyunsaturated oil	1¾ fl. oz
60 ml	Egg whites	2 egg whites
150 ml	Cold water	5 fl. oz
5 g	Salt	1 tspn

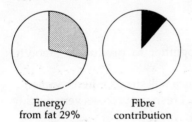

Energy
from fat 29%

Fibre
contribution

BASIC METHOD: PLAIN PASTA

1 Sieve together the plain flour and the besan flour.
2 Thoroughly mix all the ingredients to produce a smooth, pliable
   dough, then cover and allow to rest for 1 hour.
3 Shape the pasta as desired, cook in boiling water for 3–4 minutes. See
   pasta dishes on page 95.

BASIC METHOD: WHOLEMEAL PASTA

Start at step 2 and repeat.

FRUIT STRUDELS

Different types of fruit strudel are easily made by changing the fruit in the filling, for example cherries, plums and apricots.

Apple strudel

10 portions

Each portion contains:
879 kJ/207 kcal, 4 g fibre,
4 g fat.

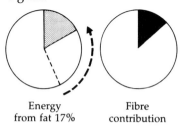

Energy from fat 17%   Fibre contribution

Dough		
200 g	Bread flour	7 oz
100 g	Wholemeal flour	3½ oz
60 ml	Egg whites	2 egg whites
30 ml	Polyunsaturated oil	1 fl. oz
80 ml	Water	3 fl. oz approx.

Filling		
800 g	Small pieces of eating apples	1 lb 12 oz
120 g	Sultanas	4 oz
5 g	Cinnamon	1 tspn

A little polyunsaturated oil to brush over the strudel (5 ml/1 tspn) and a very small quantity of icing sugar to dust over the strudel for service (5–10 g)

METHOD

1 Prepare the dough: mix the two flours together, add the rest of the ingredients and dough up. Work the dough until smooth and pliable, then allow to rest for 1–2 hours.
2 Prepare the ingredients for the filling and mix together.
3 Roll out the dough until very thin, place onto a floured cloth.
4 Carefully pull the paste apart with the back of the hands until very very thin.
5 Place the filling on the prepared dough, leaving a short edge at the sides and a top edge of 100 mm/4 inches along the length of the dough.
6 With the aid of the cloth, roll up the dough and filling up to the top edge.
7 Lightly brush over the surface with oil then roll over the top edge—this will be the top of the strudel.
8 Lift the cloth and strudel and roll onto a lightly greased and floured baking tray with the top surface upwards.

continued overleaf

9 Lightly brush over the surface with oil.
10 Bake at 200°C/400°F until cooked and golden brown, 40 minutes approximately.
11 Remove from the oven and lightly coat the surface with icing sugar.
12 Cut into portions and accompany with 200 ml/7 fl. oz Chantilly cream (see page 217).

SPONGE AND SPONGE PRODUCTS

Sponges usually consist only of plain flour, eggs and sugar. The following recipe contains less than half the eggs of a traditional sponge and is therefore not only lower in fat but also very economical.

The quality of this sponge makes it suitable for all dishes containing sponges and most cake recipes.

Cookery hints

- Always carefully blend the ingredients together; heavy handling means low volume.
- Use a polyunsaturated white fat to lightly grease the cake tin, then coat with a thin film of flour. Never use butter or margarine because the sponge is likely to stick to the cake tin.
- Ensure the sponge is cooked before removing it from the oven. When in doubt, a small skewer or cocktail stick may be inserted into the centre of the sponge; when withdrawn the stick should be clean with no sign of uncooked mixture.
- Do not forget to add glycerine to the recipe or the sponge will quickly stale.

Basic recipe

14–16 portions

Each sponge contains:
12,343 kJ/2903 kcal,
13 g fibre, 36 g fat.

Energy
from fat 11%

Batter		
130 g	Castor sugar	4½ oz
250 g	High ratio cake flour	9 oz
250 ml	Skimmed milk (colour if desired)	9 oz
15 g	Glycerine	½ oz
150 ml	Eggs (3)	5 oz
150 g	Castor sugar	5 oz
130 g	High ratio cake flour	4½ oz
20 g	Baking powder	2 dspns

METHOD

1 Prepare the batter: sieve the dry ingredients into a mixing bowl, then slowly blend in the milk and glycerine and whisk to a smooth batter.
2 Warm the eggs (32°C/90°F), place into a mixing bowl with the sugar and whisk at medium speed until very thick (ribbon stage), 15–20 minutes.
3 Add the flour batter to the whisked egg mixture and carefully blend together by hand.
4 Sieve together the flour and baking powder, then carefully blend by hand into the egg mixture.
5 Bake at 190°C/375°F for 25 minutes approximately.

Chocolate/ carob sponge

Prepare as above but replace 50 g/1½ oz of the flour with cocoa powder or carob powder.

Coffee sponge

Prepare as plain sponge but thoroughly dissolve 30 g/1 oz coffee extract in the milk before use.

Victoria sandwich

16 portions
Recipe using a 250 mm diameter and 40 mm deep (9¾ inch diameter and 1½ inch deep) cake tin.

Each portion contains: 855 kJ/204 kcal, 1 g fibre, 2 g fat.

	Recipe for low-fat sponge, page 199.	
120 g	Raspberry jam	4 oz
10 g	Icing sugar for lightly dusting over the surface	1 dspn

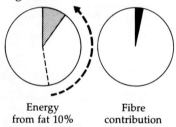

Energy   Fibre
from fat 10%   contribution

METHOD

1 Bake the sponge as described in basic low-fat sponge recipe, turn out onto a cooling wire and allow to cool.
2 Split the sponge through the centre, then sandwich the two halves together with the jam.
3 Lightly dredge over the top with icing sugar.

Eve's pudding

16 portions

Each portion contains: 965 kJ/230 kcal, 2 g fibre, 2 g fat.

	Recipe for low-fat sponge, page 199.	
1 kg	Peeled sliced cooking apples	2 lb 3 oz
90 g	Castor sugar	3 oz
Pinch	Ground cloves	Pinch
30 ml	Water	1 fl. oz

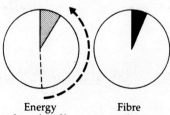

Energy   Fibre
from fat 9%   contribution

METHOD

1 Prepare the sponge as in basic recipe.
2 Mix together the apples, sugar, cloves and water, then place into a suitable pie dish.
3 Spread the sponge mixture neatly across the top.
4 Bake at 190°C/375°F for 20 minutes then reduce the temperature to 180°C/350°F and complete the cooking—a further 25 minutes approximately. *Ensure the sponge is cooked before removing from the oven. If necessary cover with a sheet of damp greaseproof paper to avoid excessive colour development.*
5 Accompany with 500 ml/18 fl. oz skimmed milk custard sauce (page 215).

Fruit trifle

10 portions

Each portion contains:
1167 kJ/278 kcal, 2 g fibre, 8 g fat.

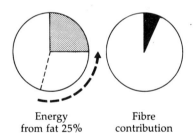

Energy from fat 25%	Fibre contribution

see overleaf for method

400 g	Low-fat sponge (page 199)	14 oz
60 g	Raspberry jam	2 oz
500 ml	*Skimmed milk custard sauce (thick)	18 fl. oz
280 ml	Whipped cream (page 217)	½ pt
300 g	Cubes or small slices of poached or tinned, lightly sweetened fruit, e.g. pears, peaches, etc.	10½ oz
90 ml	Unsweetened fruit juice (to moisten the sponge)	3 fl. oz

Decoration

5	Glacé cherries (30 g)	5
10 pieces	Angelica (5 g)	10 pieces
10 pieces	Flaked almonds (5 g)	10 pieces

Custard sauce

500 ml	Skimmed milk	18 fl. oz
30 g	Custard powder	1 oz
30 g	Sugar	1 oz

METHOD

1 Split the sponge through the centre then sandwich together with the jam. Cut into a neat dice, 15 mm/½ inch approximately.
2 Place into the serving dish with the fruit then moisten with the fruit juice.
3 Prepare the custard sauce (page 215) and pour over the sponge. Cool quickly and thoroughly chill.
4 Pipe the cream over the surface of the custard and decorate with the cherries, angelica and flake almonds.

Steamed sponge puddings

16 portions

	Basic recipe for low-fat sponge, page 199.	
800 ml	Skimmed milk custard sauce (or any other suitable sauce) (see page 215)	1 pt 8 fl. oz

METHOD

1 Place the basic mixture into lightly greased pudding moulds and cover with lightly greased greaseproof paper.
2 Steam for 1½ hours (depending on size).
3 Turn out onto a serving dish and serve with the sauce.

Vanilla sponge pudding

(16 portions)

Add a few drops of vanilla essence to the basic sponge and serve with a skimmed milk vanilla-flavoured sauce—500 ml/18 fl. oz.

Currant, raisin or sultana sponge pudding

16 portions

Each portion contains:
851 kJ/203 kcal, 2 g fibre,
2 g fat.

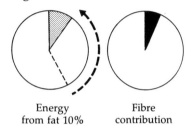

Energy
from fat 10%

Fibre
contribution

Add 120 g/4 oz selected fruit (washed and thoroughly dried) to the basic sponge.

GATEAUX

Chocolate gâteau

16 portions

Recipe using a 250 mm diameter and 40 mm deep (9¾ inch diameter and 1½ inch deep) cake tin.

Each portion contains:
1070 kJ/253 kcal, 1 g fibre,
8 g fat.

	Chocolate sponge (basic recipe, page 199)	
250 ml	Whipped cream (page 217)	9 fl. oz
60 g	Chocolate vermicelli	2 oz
30 g	Grated chocolate	1 oz
120 ml	Unsweetened fruit juice (to moisten the sponge)	4 fl. oz

See overleaf for method

Energy
from fat 27%

Fibre
contribution

METHOD

1   Split the sponge through the centre then dampen with the fruit juice—*this is necessary to produce a gateau with a moist eating texture.*
2   Sandwich together the two pieces of sponge with two-thirds of the whipped cream, apply a thin coat of cream around the sides.
3   Cover the sides with the vermicelli.
4   Spread the remaining whipped cream over the top of the gateau then neatly mark with a serrated scraper.
5   Sprinkle the grated chocolate over the surface of the gateau.

Coffee gâteau

16 portions

Recipe using a 250 mm diameter and 40 mm deep (9¾ inch diameter and 1½ inch deep) cake tin.

Each portion contains: 1162 kJ/268 kcal, 2 g fibre, 10 g fat.

	Coffee sponge (basic recipe, page 199)	
250 ml	Whipped cream (page 217)	9 fl. oz
90 g	Nib almonds	3 oz
60 g	Toasted flaked almonds	2 oz
120 ml	Unsweetened fruit juice	4 fl. oz

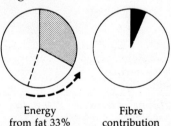

Energy
from fat 33%

Fibre
contribution

METHOD

Prepare the same as the chocolate gateau above using coffee sponge and almonds in place of chocolate sponge and chocolate garnish. Coat the sides of the sponge with the nib almonds and decorate the top with the toasted flaked almonds.

Fruit gâteau

16 portions

Recipe using a 250 mm diameter and 40 mm deep (9¾ inch diameter and 1½ inch deep) cake tin.

Each portion contains: 1209 kJ/287 kcal, 2 g fibre, 10 g fat.

	Vanilla sponge (basic recipe, page 199)	
250 ml	Whipped cream (page 217)	9 fl. oz
90 g	Toasted flaked almonds	3 oz
120 ml	Unsweetened fruit juice	4 fl. oz
300 g	Cubes or slices of poached or tinned lightly sweetened fruit (or appropriate fresh fruit)	10½ oz

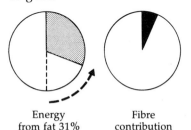

Energy from fat 31%   Fibre contribution

METHOD

1 Prepare the gateau as previously stated, filling the centre with half the quantity of whipped cream and half the quantity of fruit.
2 Coat the sides with the toasted flaked almonds and decorate the top with the remaining cream and fruit.

Apricot cheese gateaux

16 portions

	Basic sponge recipe (page 199)	
100 g	Dried apricots	3½ oz
600 g	Apricot/fromage frais	1 lb 5 oz

METHOD

1 Poach 16 of the apricots and reserve the juice to moisten the gateaux, and the fruits for decoration. Finely dice the remaining apricots.
2 Prepare the gateaux as stated in the basic recipe, adding the diced apricots at point 4.
3 Slice the gateaux into two or three layers, moisten with the juices and apricot liquer (optional). Layer with half the fromage frais and use the remainder to decorate the sides and top.

NAN BREADS AND PIZZAS

NAN BREAD

(10 portions)

A special Indian bread suitable for many Eastern dishes, including Tandoori chicken on page 159. The bread is usually cooked in a special clay oven called a 'tandoor', but a hot frying pan and grill make ideal substitutes. This bread may be stored in a deep freeze after cooking and used as and when required.

Plain nan bread

10 portions

Each portion contains:
977 kJ/230 kcal, 2 g fibre, 4 g fat.

Energy   Fibre
from fat 16%   contribution

500 g	Plain flour	1 lb 2 oz
5 g	Salt	1 tspn
10 g	Baking powder	2 tspns
15 g	Castor sugar	½ oz
150 ml	Yogurt	5 fl. oz
30 ml	Polyunsaturated oil	1 fl. oz
30 g	Fresh yeast	1 oz
170 ml	Skimmed milk (room temperature)	6 fl. oz
10 g	Sesame seeds	2 tspns

Wholemeal nan bread

10 portions

Each portion contains:
911 kJ/215 kcal, 5 g fibre, 5 g fat.

Energy   Fibre
from fat 19%   contribution

500 g	Wholemeal flour	1 lb 2 oz
5 g	Salt	1 tspn
10 g	Baking powder	2 tspns
15 g	Castor sugar	½ oz
150 ml	Yogurt	5 fl. oz
30 ml	Polyunsaturated oil	1 fl. oz
30 g	Fresh yeast	1 oz
200 ml	Skimmed milk (room temperature)	7 fl. oz
10 g	Sesame seeds	2 tspns

BASIC METHOD

1 Thoroughly mix together the dry ingredients, that is, the flour, salt, baking powder and sugar.
2 Place the yeast in a small bowl, add a little of the milk and blend together.
3 Add the yeast mixture, yogurt, oil and milk to the flour and mix to a smooth dough.
4 Cover and allow to rest at room temperature for 15 minutes.
5 Divide the dough into 10 pieces, then roll out thinly using flour and sesame seeds (traditionally the dough is shaped like a large teardrop).
6 Place the shapes onto oiled plates and allow to dry prove (ferment) at room temperature for 30–40 minutes.
7 Carefully turn each nan out of the plate and into a hot, very lightly greased frying pan and cook until the base is brown (30 seconds approximately).
8 Place under a very hot grill and brown.
9 Serve hot.

PIZZAS
(10 portions)

The following recipe for the pizza base is designed to produce good yeast pizzas with very little handling and a short fermentation time. For nutritional analysis of plain dough, see Appendix 2.

Plain pizza dough

500 g	Plain flour	18 oz
5 g	Salt	1 tspn
30 ml	Polyunsaturated oil	1 fl. oz
30 g	Fresh yeast	1 oz
300 ml	Skimmed milk (room temperature)	½ pt

BASIC METHOD

1 Sieve together the salt and flour.
2 Blend the yeast with a little of the milk then add to the flour.
3 Add the remaining milk and oil and mix to a smooth pliable dough.
4 Cover and allow to ferment at room temperature for 15 minutes.
continued overleaf

5 Divide the dough into portions then mould into balls.
6 Roll out the balls of dough into rounds and brush over the surface with a little oil.
7 Add the topping, leaving a 10 mm/⅜ inch clear border round the pie.
8 Allow to dry prove (ferment) at room temperature for 30–40 minutes.
9 Bake at 225°C/430°F for 15–20 minutes.

Wholemeal pizza dough

As above but replace the plain flour with wholemeal flour and increase the skimmed milk to 350 ml/12 fl. oz.

Tomato and vegetable pizza

10 portions

Each portion contains: '
1236 kJ/293 kcal, 3 g fibre,
8 g fat.

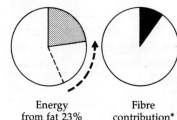

Energy from fat 23% · Fibre contribution*

*Double if wholemeal dough is used

850 g	Pizza dough (plain or wholemeal)	1 lb 14 oz
120 g	Chopped onion	4 oz
5 g	Crushed garlic	2 tspns
600 g	Tinned tomatoes	1 lb 5 oz
15 g	Tomato purée	½ oz
2 g	Oregano	1 dspn
90 g	Sliced mushrooms	3 oz
90 g	Diced peppers	3 oz
90 g	Sweetcorn kernels	3 oz
200 g	Low-fat cheddar-type cheese	7 oz
10 ml	Polyunsaturated oil (for brushing over top of pizzas)	2 tspns

PREPARATION OF FILLING

1 Lightly cook the onions, garlic, tinned tomatoes, tomato purée, and oregano in a saucepan.
2 Add the peppers, mushrooms and sweetcorn and continue cooking for a short period; *but keep crisp and do not overcook*.
3 Allow to cool and use when required.
4 Complete as stated above.

SCONES

This is an excellent way of preparing scones with a lower fat content than traditional scones.

Cookery hints

- The dough will be very soft and sticky, therefore use a good base of flour to roll out the mixture. However, avoid additional flour being absorbed into the mixture.
- Use self-raising flour when rolling out the dough; this is important to ensure that the trimmings of dough will still produce good scones.
- Handle the dough lightly—too much handling produces tough low volume scones.
- Bake in a hot oven and do not over-cook. Even short periods of over-cooking will produce dry eating scones.
- When commercial baking powder is used (not cream of tartar type), leave the scones to rest for 15 minutes prior to baking. This allows the scones to rest after handling.

Plain scones

8 scones

Each portion contains:
666 kJ/157 kcal, 1 g fibre, 4 g fat.

30 ml	Polyunsaturated oil	1 fl. oz
30 g	Castor sugar	1 oz
30 ml	Egg white (1 egg white)	1 fl. oz
120 ml (scant)	Skimmed milk	4 oz
225 g	*Self-raising flour	8 oz

*To make self-raising flour sieve together 5 g baking powder per 100 g plain flour.

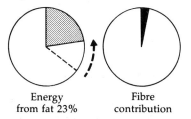

Energy from fat 23%   Fibre contribution

METHOD

1 Whisk together the oil, sugar, egg white and milk.
2 Add the flour and lightly mix to a dough.
3 Scrape out the mixture onto a board dusted with self-raising flour.
4 Dust over the top with flour, then roll out to 25 mm/1 inch thickness.
5 Cut out the scones with a plain cutter and place onto a lightly greased baking tray.

6   Leave plain or brush with egg wash and lightly stab with a fork.
7   Allow to rest for 15 minutes (see cookery hints above) then bake at
    240°C/450°F for 10–15 minutes.

Fruit scones   (sultana, raisin, currant)
Add 60 g/2 oz appropriate fruit to the basic recipe.

Wholemeal scones
Replace the flour in the basic recipe with wholemeal flour mixed with
baking powder (5 g baking powder per 100 g flour) and increase the milk
to 150 ml/5 fl. oz.

Fruit pizzas

10 portions

*An unusual addition to the pizza
scene.*

Each portion contains:
756 kJ/178 kcal, 1 g fibre,
4 g fat.

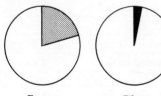

Energy
from fat 21%

Fibre
contribution

Pizza base		
400 g	Basic scone dough (see recipe on page 209)	14 oz

Fruit topping		
500 g	Prepared fruit or tinned fruit, lightly sweetened, e.g. pears, peaches, cherries, etc.	1 lb 2 oz
10 ml	Polyunsaturated oil for lightly brushing the fruit	2 tspns
75 ml	Fruit glaze for brushing over the pizzas when baked	2½ fl. oz

METHOD

1   Prepare the scone dough as stated on page 209.
2   Roll out into a round pizza shape, 10 mm/½ inch in thickness.
3   Lightly brush over the surface of the dough with a little of the oil to
    reduce moisture absorption from the fruit.
4   Drain the fruit well and wipe off excess surface moisture. Leave small
    fruits whole and cut large fruits into suitable pieces or slices.
5   Neatly arrange the fruit over the top of the dough, lightly brush with a
    little oil.
6   Bake in a hot (230°C/450°F) oven until cooked, 12–15 minutes.
7   Coat the fruit with the hot fruit glaze, serve hot or cold.

FRESH FRUIT KEBABS

A novel and interesting sweet which may be made with all types of fruit, for example grapes, kiwi fruit, apples, mangoes, plums, cherries and strawberries.

Cookery hint

- Most fruits only require heating through, therefore avoid over-cooking as the fruit will break up and drop off the skewer.

Recipe

10 portions

Each portion contains:
371 kJ/87 kcal, 4 g fibre,
1 g fat.

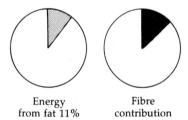

| Energy from fat 11% | Fibre contribution |

1700 g	Fresh fruit, to taste, e.g. 2 bananas—peeled cut into 20 mm thick pieces 2—peaches cut into pieces 2 apples—cut into pieces 2 pears—cut into pieces 10 pineapple cubes (150 g) 2 oranges—cut into segments	3 lb 12 oz
15 ml	Lemon juice (to prevent discoloration)	½ fl. oz
	Mint leaves (optional)	
10 ml	Polyunsaturated oil	2 tspns

METHOD

1 Arrange the fruit (and mint leaves if used) neatly on the skewers, lightly brush over the surfaces with the oil.
2 Place the skewers on a tray, lightly grill until heated through. Turn once during cooking and lightly baste the surfaces of the fruit with oil.

Note: For high-class service the kebabs may be flamed with brandy when serving.

STARCH PUDDINGS

(rice, semolina and mixed cereal)

Cookery hints

- Rinse the cooking pan with water before adding the milk; this reduces the risk of burning the base of the pan.
- Add the sugar at the end of the cooking period; this also reduces the risk of burning the pudding.

Basic recipe

10 portions

1¼ litres	Skimmed milk	2 pt 4 fl. oz
80 g	Castor sugar	2½ oz
To taste	Vanilla essence	To taste

METHOD

1 Place the milk in a saucepan and bring to the boil.
2 Sprinkle in the main item/s, stirring constantly to avoid forming lumps.
3 Slowly simmer until cooked, stirring frequently.
4 Add the sugar and vanilla essence, stir until the sugar is dissolved.

Serve in a pie dish, earthenware dish or any other suitable dish: pour the mixture into the dish, then brown the surface under a hot grill.

Rice pudding

10 portions

Each portion contains:
496 kJ/116 kcal, <1 g fibre, <1 g fat.

	Basic recipe (above)	
120 g	Round grain rice	4 oz
Pinch	Nutmeg	Pinch

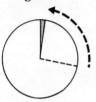

Energy
from fat 2%

METHOD

Follow the instructions in the basic recipe. When the rice has been transferred to the pie dish, sprinkle over the surface with the nutmeg.

Semolina, sago and tapioca pudding

Basic recipe (opposite)		
100 g	Main item	3½ oz

METHOD

Follow the instructions in the basic recipe.

Mixed pudding

10 portions

Each portion contains:
554 kJ/134 kcal, 1 g fibre,
<1 g fat.

	Basic recipe (opposite)	
60 g	Round grain rice	2 oz
30 g	Brown rice	1 oz
30 g	Millet	1 oz
60 g	Raisins	2 oz

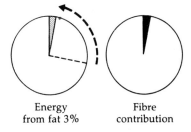

Energy
from fat 3%

Fibre
contribution

METHOD

Follow the instructions in the basic recipe, adding the raisins at the end of the cooking period.

Bread pudding

10 portions

Each portion contains:
695 kJ/165 kcal, 2 g fibre,
6 g fat.

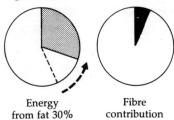

Energy
from fat 30%

Fibre
contribution

6 slices	Wholemeal bread (crusts removed)	6 slices
30 g	Polyunsaturated margarine	1 oz
60 g	Sultanas	2 oz
800 ml	Skimmed milk	1 pt 8 fl. oz
250 ml	Eggs	5 eggs (size 3)
60 g	Castor sugar	2 oz
To taste	Vanilla essence	To taste
5 g	Icing sugar	2 tspns

METHOD

1 Place the eggs, sugar and vanilla essence into a bowl, whisk together.
2 Heat the milk, whisk into the egg and sugar.
3 Spread the slices of bread with the margarine.
4 Place the bread and sultanas into a pie dish.
5 Strain the mixture over the bread and sultanas and allow to stand until the bread absorbs the custard mixture.
6 Place the dish into a tray half full of water, cook in a cool oven, 160°C/320°F.
7 When cooked, lightly dust over the surface with icing sugar.

Note: Make sure that none of the sultanas remain on the surface of the pudding, because they will burn and spoil the flavour.

SWEET SAUCES

Cornflour thickened sauces

1 litre contains:
3,031 kJ/707 kcal, 1 g fibre, 4 g fat.

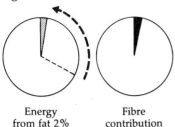

Energy from fat 2%

Fibre contribution

Basic recipe
1 litre/1¾ pt

1 litre	Skimmed milk	1¾ pt
40 g	Cornflour	1½ oz
60 g	Sugar	2 oz

METHOD

1 Place a little of the cold milk into a bowl, add the cornflour and mix together.
2 Place the remaining milk into a saucepan, bring to the boil.
3 Stir the diluted cornflour mixture into the hot milk. Stir constantly until a smooth sauce is obtained.
4 Reboil and simmer for 2–3 minutes, stirring occasionally to avoid burning.
5 Remove from the boil and mix in the recipe sugar and any essence, spirit or wine.

Almond sauce   *(1 litre/1¾ pt)*
Basic recipe, adding 4–5 drops almond essence.

Brandy, whisky and sherry sauces   *(1 litre/1¾ pt)*
Basic recipe, adding 60 ml/2 fl. oz appropriate spirit or wine.

Chocolate sauce   *(1 litre/1¾ pt)*
Basic recipe, adding 60 g/2 oz cocoa powder. Blend the cocoa powder with the milk and cornflour before thickening the sauce.

Custard sauce   *(1 litre/1¾ pt)*
Basic recipe using custard powder instead of cornflour.

FRUIT SAUCES *(1 litre/1¾ pt)*

Cookery hints

- A little lemon juice will enhance the flavour of less acidic fruits, for example strawberries, mangoes, paw paw and melon.
- The quantity of arrowroot required to thicken the sauce will depend on the type of fruit used and whether the sauce is to be served hot or cold. Use less arrowroot for thick fruit purées and cold sauces.

Basic recipe

for raspberry (Melba), strawberry, cherry, plum, etc.

1 litre contains:
1440 kJ/340 kcal, <1 g fibre, trace of fat.

900 g	Tinned unsweetened fruit (with juice)	2 lb
30 g	Arrowroot	1 oz
To taste	Lightly sweeten	To taste

METHOD

1 Place a little of the cold fruit juice into a bowl, add the arrowroot and mix together.
2 Purée the fruit and juice in a food processor or liquidizer, place in a saucepan and bring to the boil.
3 Stir in the diluted arrowroot in stages until the desired thickness is obtained.

WHIPPED CREAM FOR PIPING AND SAUCES

Cookery hints

- Always lightly whisk cream until thick when serving as a sauce—see next hint.
- Always use whipping cream (35% butterfat). This not only contains less fat than double cream but also traps more air; provided the cream is not over-whipped and a small quantity of sugar is used, the cream should almost double in volume. Because of this increase in volume, all recipes which contain cream have volume measurements (not weight).
- When serving cream always use a teaspoon. This should result in small portions being taken or served.

Recipe for whipped cream

1 litre/1¾ pt.

Each litre contains:
8432 kJ/2043 kcal, 203 g fat.

580 ml	Whipping cream (35% butterfat)	1 pt
30 g	Castor sugar	1 oz

Energy
from fat 89%

Whip the chilled cream and sugar to full volume then use as required.

Chantilly cream
Prepare as above, adding vanilla essence to taste.

GELATINE-BASED SWEETS

Gelatine may be used as a foundation for a large number of sweets including cold mousses, Bavarian cream, charlottes and cold summer puddings. To be creative, try using the mousse mixture as a filling for gateaux and flans and also as a topping, for example, for trifles and jellies.

COLD MOUSSES

Cookery hints

- To produce the very light texture of a good mousse, ensure the mixture is almost setting before carefully folding through the cream and egg whites.
- Rinsing the mould with cold water prior to adding the mixture reduces the likelihood of the mousse sticking when demoulding.

Vanilla mousse

10 portions

Each portion contains:
442 kJ/104 kcal, no fibre,
2 g fat.

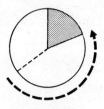

Energy
from fat 19%

1140 ml	Skimmed milk	2 pt
70 g	Castor sugar	2½ oz
20 g	Cornflour	2 dspns
30 g	Leaf gelatine	1 oz
60 ml	Whipping cream	2 fl. oz
2½ ml	Vanilla essence	4–5 drops
50 ml	Egg whites	2 egg whites

METHOD

1 Blend the cornflour with a little of the cold milk.
2 Place the remaining milk in a saucepan, bring to the boil.
3 Stir the diluted cornflour mixture into the hot milk.
4 Reboil and simmer for 1 minute, stirring occasionally to avoid burning.
5 Add the sugar.

6 Place the leaf gelatine in a bowl of cold water, allow to soften.
7 Squeeze out surplus water from the gelatine, add to the mixture and stir until completely dissolved.
8 Pass the mixture through a fine strainer into a clean bowl.
9 Cool quickly, place the bowl on ice and stir to almost setting point.
10 Meanwhile, lightly whip the cream and whisk the egg whites to a stiff snow.
11 Carefully fold the cream through the setting mixture, followed by the egg whites.
12 Pour the mixture into a suitable mould, place in a refrigerator and allow to set.
13 When set, dip the mould in tepid water and demould onto a serving dish.
14 Decorate the mousse to taste with slices of fruit or red cherries, etc.

Chocolate mousse

10 portions

Each portion contains:
520 kJ/123 kcal, no fibre,
4 g fat.

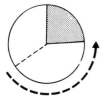

Energy
from fat 24%

Prepare as above, omitting the vanilla essence and add 60 g/2 oz cocoa powder. Blend together the cocoa powder with a little of the cold milk. Stir into the hot milk until combined, then complete as stated.

Carob mousse (chocolate alternative)
Prepare as above using carob powder* instead of cocoa powder.

*Carob powder contains only 1% fat.

Fruit mousses

Prepare as vanilla mousse but reduce the milk to 700 ml/1¼ pt and omit the vanilla essence. Add 450 g/1 lb unsweetened fruit purée, for example raspberries, strawberries, peaches or pineapple*, etc.

Orange mousse

Add the grated zest of 4 oranges into the milk instead of the vanilla essence.

Lemon mousse

Add the grated zest of 4 lemons into the milk instead of the vanilla essence.

Charlotte russe

10 portions

Each portion contains:
789 kJ/186 kcal, <1 g fibre, 3 g fat.

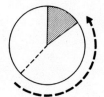

Energy
from fat 15%

	Basic vanilla mousse recipe (see page 218)	
200 g	Sponge (see page 199)	7 oz
150 ml	Raspberry table jelly	¼ pt

METHOD

1 Set the jelly in the bottom of a large charlotte mould or any other suitable mould.
2 Cut the sponge into finger shapes then line the sides of the mould.
3 When the mousse has been prepared, pour into the lined mould and allow to set in a refrigerator.
4 When set, demould as vanilla mousse, page 218.

*Heat treat the pineapple purée to destroy the enzyme bromylin.

Charlotte montreuil

10 portions

Each portion contains:
796 kJ/187 kcal, <1 g fibre,
3 g fat.

Energy
from fat 14%

	Basic mousse recipe with unsweetened peach purée (page 218)	
200 g	Sponge (page 199)	7 oz
120 g	Sliced peaches (tinned/unsweetened)	4 oz

METHOD

1 Cut the sponge into finger shapes then line the sides of a large charlotte mould.
2 Pour the mousse into the mould, allow to set.
3 Demould, then decorate the top of the charlotte with the sliced peaches.

Charlotte andalouse

Prepare the same as charlotte montreuil above, using an orange flavoured mousse and decorating the top of the charlotte with orange segments.

COLD RICE DISHES

Impératrice rice

10 portions

Each portion contains:
685 kJ/161 kcal, <1 g fibre,
3 g fat.

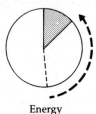

Energy
from fat 13%

1¼ litres	Skimmed milk	2 pt 4 fl. oz
80 g	Castor sugar	2½ oz
60 g	Round grain rice	2 oz
60 g	Brown rice	2 oz
2½ ml	Vanilla essence	4–5 drops
15 g	Leaf gelatine	½ oz
60 ml	Whipping cream	2 fl. oz
50 ml	Egg whites	2 egg whites
150 ml	Red table jelly	¼ pt
30 g	Diced glacé cherries	1 oz
15 g	Diced angelica	½ oz

METHOD

1 Set the jelly in a large charlotte mould.
2 Prepare a thin milk pudding with the milk, rice, vanilla essence and sugar—see basic recipe for milk puddings, page 212.
3 Soak the leaf gelatine in a bowl of cold water until soft and pliable.
4 Squeeze out the surplus water from the gelatine, add to the hot milk pudding and stir until completely dissolved.
5 Allow to cool then place over ice and stir to almost setting point.
6 Meanwhile, lightly whip the cream and whisk the egg whites until stiff.
7 Add the glacé cherries and angelica to the mixture then fold through the cream followed by the stiffly whisked egg whites.
8 Allow to set then demould the same as vanilla mousse, page 218.

Palermitaine rice

Prepare the same as above but set in a savarin mould. Fill the centre with strawberries (300 g/10½ oz).

Pineapple créole

10 portions

Prepare the rice the same as impératrice but omit the jelly, glacé cherries and angelica. When the rice is well chilled and almost set, dress in the shape of a half pineapple on a serving dish and allow to set. Arrange slices of pineapple (450 g/1 lb) over the top of the rice to resemble a half pineapple. Coat with a little unsweetened pineapple glaze.

Cheesecake

10 portions

(225 mm/9 inch diameter cake ring/separating cake tin)

Each portion contains:
1047 kJ/249 kcal, no fibre,
9 g fat.

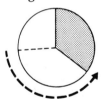

Energy
from fat 33%

400 ml	Skimmed milk	14 fl. oz
100 g	*Skimmed milk powder	3½ oz
60 g	**Castor sugar	2 oz
10 g	Cornflour	1 dspn
10 g	Leaf gelatine	4 sheets
400 g	Fromage frais or quark	14 oz
30 ml	Lemon juice	1 fl. oz
150 g	Cream cheese	5 oz
50 ml	Egg whites	2 egg whites
200 g	Low-fat sponge (page 199) (1 circle of sponge 225 mm/ 9 inches)	7 oz

METHOD

1 Cut a circle of sponge to fit the base of a 225 mm/9 inch diameter cake or flan ring. Use the cake ring as a guide to cut the sponge the exact shape.
2 Place the cake ring with the sponge on top of a serving dish or cake board. Also line a band of lightly oiled paper around the inside of the cake ring.
3 Place a little of the cold milk into a bowl, add the cornflour and mix together.
4 Place the remaining milk into a saucepan and whisk in the milk powder.*

continued overleaf

*Check whether the milk powder has to be added to cold or hot milk.
**A little saccharin (4 small tablets) may be added to the hot milk if a sweeter mixture is desired.

5   Bring to the boil, stir in the diluted cornflour mixture.
6   Reboil and simmer for 1 minute, stirring occasionally to avoid burning.
7   Add the sugar.
8   Meanwhile, place the leaf gelatine into a bowl of cold water and allow to soften.
9   Squeeze out the water from the gelatine, add to the mixture and stir until completely dissolved.
10 Pass the mixture through a fine strainer into a clean bowl.
11 Meanwhile cream together the two cheeses and the lemon juice. Also whisk the egg whites to a stiff snow.
12 Cool the mixture then add the cheese and lemon mixture.
13 Place the bowl on ice and stir to almost setting point.
14 Fold through the stiffly whisked egg whites.
15 Pour the mixture into the cake ring, place in a refrigerator and allow to set.
16 When set, carefully remove the cake ring and lining paper.

Fruit cheesecakes   (10 portions)
Decorate the top of the cheesecake with 450 g/1 lb fresh or unsweetened tinned fruit, for example strawberries, raspberries, black cherries, peaches, sliced pineapple, blackcurrants, etc. Lightly coat with a little fruit glaze (low-sugar fruit syrup thickened with arrowroot).

REDUCED FAT ICE-CREAM

Freshly made ice-creams are delicious and easy to prepare provided an ice-cream machine is available. One disadvantage is that they do not store as well as commercially prepared ice-cream and suffer from loss of texture during storage. However, the use of glycerine improves the keeping properties of the ice-cream for short periods.

Ice-creams prepared on the premises must be made in a hygienic manner and should be served as fresh as possible.

Important: The production of ice-cream for consumption by members of the general public within the UK is governed by legislation, therefore it is advisable to consult the nearest local authority on production and service to the public.

Vanilla ice-cream

Cream based ice-cream

40 portions

Each portion contains:
324 kJ/77 kcal, no fibre,
3 g fat.

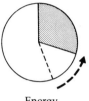

Energy
from fat 35%

2 litres	Whole fresh milk	3½ pt
100 g	Dried skimmed milk powder	3½ oz
50 g	Cornflour	1¾ oz
100 ml	Cold milk	3½ fl. oz
200 g	Sugar	7 oz
20 ml	Vanilla essence	½ fl. oz
100 ml	Single cream	3½ fl. oz
20 g	Glycerine	½ fl. oz

METHOD

1 Place the milk in a saucepan then whisk in the milk powder.*
2 Bring to the boil.
3 Blend together the cornflour and cold milk to form a smooth paste.
4 Stir the diluted cornflour mixture into the hot milk.
5 Reboil and simmer for 1 minute, stirring occasionally to avoid burning.
6 Add the sugar and glycerine, cool quickly.
7 Add the cream and vanilla essence then freeze in an ice-cream machine.

FRUIT ICE-CREAMS

Add 400 g/14 oz prepared fruit purée to the ice-cream mixture. This can easily be done by liquidizing the prepared fruit with a little of the cold ice-cream mixture.

Raspberry ice-cream
Add 400 g/14 oz raspberries to mixture.

Strawberry ice-cream
Add 400 g/14 oz strawberries.

*Check whether the milk powder has to be added to cold or hot milk.

COUPES   *(Recipes per individual portion)*

Coupes usually consist of ice-cream and fresh or poached fruit served in a coupe dish with a topping of fruit sauce and/or whipped cream. Although the following coupes are all traditional recipes, it is worthwhile exploring combinations of flavours and textures; using fruits or fruit ice-creams such as mangoes, paw paw and lychees with spiced sauces (fruit sauces flavoured with root ginger, lime, etc., see page 216).

Note: The calculations are based on 1 portion of the relevant ice-cream (page 225)

Alexandra

Each portion contains:
469 kJ/112 kcal, 1 g fibre,
3 g fat.

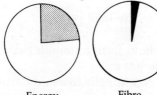

Energy from fat 24%   Fibre contribution

	Recipe for strawberry ice-cream (page 225)	
60 g	Fruit salad (fresh, unsweetened tinned)	2 oz
2½ ml	Kirsch (optional)	level tspn

Place the fruit salad (fresh/unsweetened tinned) in a coupe and top with a ball of strawberry ice-cream. A little kirsch may be added to the fruit salad.

Edna-May

Each portion contains:
458 kJ/109 kcal, 1 g fibre,
3 g fat.

	Recipe for vanilla ice-cream (page 225)	
60 g	Cherries (ripe or poached)	2 oz
10 ml	Melba sauce (page 216)	1 tblspn

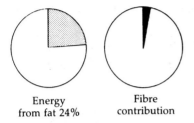

Energy
from fat 24%

Fibre
contribution

Place the ripe or poached cherries in a coupe dish and top with a ball of vanilla ice-cream. Coat with the Melba sauce.

Venus

Each portion contains:
417 kJ/99 kcal, 1 g fibre,
3 g fat.

	Recipe for vanilla ice-cream (page 225)	
60 g	Peach (half)	2 oz
Decoration		
	Strawberries	

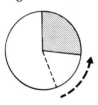

Energy
from fat 27%

Place the half peach flat side up in a coupe. Sit a ball of vanilla ice-cream on top and decorate with a large strawberry.

Glossary of terms and nutritional definitions

AMINO ACIDS
The units that make up protein chains. All amino acids have a nitrogen containing amino group (NH_2) but the rest of their structure varies with the plant or animal source. There are 20 common amino acids and eight are said to be essential as they cannot be made by the human body.

ANAEMIA
A condition which occurs because the blood cells are unable to transport sufficient oxygen around the body, usually due to a lack of iron. Anaemia causes general fatigue and listlessness and an inability to work at a normal pace.

BASAL METABOLIC RATE
The rate, or speed, at which the body uses energy for breathing, heart beat, the maintenance of body temperature and digestion. It is dependent on age and body size.

BILE SALTS
Bile salts are produced by the liver and released into the small intestine where they link with fat droplets and aid their digestion and passage into the blood.

CALCIUM
Calcium is a mineral which gives the bones strength and is essential for muscle function.

CALORIE
The unit commonly used to describe the amount of energy in food. Calorie is an abbreviation of kilocalorie (kcal) and is now often replaced by the Standard International Unit the kilojoule (kJ). 1 kcal = 4.184 kJ.

CANCER OF THE COLON
A tumour in the colon which causes an obstruction.

CARDIOVASCULAR DISEASES
Diseases of the heart and blood vessels.

CAROTENE
A yellow pigment found in vegetables and fruits which can be converted to vitamin A by the body—6 μ carotene and equivaent to 1 μ of vitamin A.

CELLULOSE
A type of indigestible carbohydrate found in plant cell walls which contributes to the fibre content of food.

CEREBROVASCULAR DISEASES
Diseases of the blood vessels in the brain.

CIRRHOSIS
A chronic disease of the liver caused by damage to the liver cells which is usually due to infection or regular heavy drinking.

COLON
The end section of the digestive system; also called the large bowl.

CONGENERS
A mixture of by-products found in fermented drinks, some of which are toxic.

CONNECTIVE TISSUE
Tissues which form the basic framework of the body.

CORONARY HEART DISEASE
A heart attack caused by failure of the blood vessels in the heart to supply sufficient oxygen to allow the heart to beat normally.

DEFICIENCY DISEASE
A disease caused by a low intake of a particular nutrient.

DENTAL CARIES
Tooth decay caused by damage to the hard outer surface of the teeth.

DIVERTICULITIS
The formation of pockets in the colon which may become infected.

ENZYMES
Proteins which link with other substances in the body and speed up their breakdown or alteration without being changed themselves. Digestive enzymes break down food into smaller units which can then pass into the blood stream, and metabolic enzymes assist in the manufacture and function of body cells.

GLYCERINE
An alternative name for glycerol.

GLYCEROL
A clear liquid with fat-like properties used to soften icing and frozen products. Glycerol is the central compound in a fat which links together the three fatty acids.

HORMONES
Substances made in the body which regulate body processes.

HYPERTENSION
An increase in blood pressure.

KILOCALORIE

The unit commonly used to describe the amount of energy in food; often abbreviated to calorie. Kilocalorie (kcal) is now being replaced by the Standard International unit, the kilojoule (kJ). 1 kcal = 4.184 kJ.

KILOJOULE

The Standard International unit used to describe the amount of energy in food (kJ).

LACTO-VEGETARIAN

A person who only eats plant foods and milk products.

MEGADOSE

An excessively large dose of a medicine.

MEGAJOULE

A megajoule (MJ) = 1000 kJ or 239 kcal.

MONOSODIUM GLUTAMATE (MSG)

A flavour enhancer which occurs naturally in seaweed and is made commercially from sugar-beet and gluten.

NITROGEN

An inert gas which makes up four-fifths of the earth's atmosphere and is an essential part of the structure of proteins.

NUTRIENT

The components of any food, liquid or solid, which can supply the body with energy or materials for growth, repair and body regulation.

PLAQUE

A sticky layer on the surface of the teeth formed by bacteria using the sugars eaten. Plaque prevents the saliva from reaching the surface of the teeth and therefore is a cause of tooth decay.

SPINA BIFIDA

A spinal defect in newborn babies.

VEGAN

A person who does not eat any foods of animal origin.

Appendix 1

Recommended daily amounts of nutrients for population groups (Department of Health and Social Security, 1979)

Age ranges years		Energy MJ	Energy kcal	Protein g	Calcium mg	Iron mg
Boys						
Under 1		3.25	780	19	600	6
1		5.0	1,200	30	600	7
2		5.75	1,400	35	600	7
3–4		6.5	1,560	39	600	8
5–6		7.25	1,740	43	600	10
7–8		8.25	1,980	49	600	10
9–11		9.5	2,280	56	700	12
12–14		11.0	2,640	66	700	12
15–17		12.0	2,880	72	600	12
Girls						
Under 1		3.0	720	18	600	6
1		4.5	1,100	27	600	7
2		5.5	1,300	32	600	7
3–4		6.25	1,500	37	600	8
5–6		7.0	1,680	42	600	10
7–8		8.0	1,900	48	600	10
9–11		8.5	2,050	51	700	12†
12–14		9.0	2,150	53	700	12†
15–17		9.0	2,150	53	600	12†
Men						
18–34	Sedentary	10.5	2,510	62	500	10
	Moderately active	12.0	2,900	72	500	10
	Very active	14.0	3,350	84	500	10
35–64	Sedentary	10.0	2,400	60	500	10
	Moderately active	11.5	2,750	69	500	10
	Very active	14.0	3,350	84	500	10
65–74		10.0	2,400	60	500	10
75 and over		9.0	2,150	54	500	10
Women						
18–54	Most occupations	9.0	2,150	54	500	12†
	Very active	10.5	2,500	62	500	12†
55–74		8.0	1,900	47	500	10
75 and over		7.0	1,680	42	500	10
Pregnant		10.0	2,400	60	1,200	13
Lactating		11.5	2,750	69	1,200	15

*Most people who go out in the sun need no dietary source of vitamin D, but children and adolescents in winter, and housebound adults, are recommended to take 10 µg vitamin D daily.   †These iron recommendations may not cover heavy menstrual losses.

Vitamin A (retinol equivalent) μg	Thiamin mg	Riboflavin mg	Niacin equivalent mg	Vitamin C mg	Vitamin D* μg
450	0.3	0.4	5	20	7.5
300	0.5	0.6	7	20	10
300	0.6	0.7	8	20	10
300	0.6	0.8	9	20	10
300	0.7	0.9	10	20	–
400	0.8	1.0	11	20	–
575	0.9	1.2	14	25	–
725	1.1	1.4	16	25	–
750	1.2	1.7	19	30	–
450	0.3	0.4	5	20	7.5
300	0.4	0.6	7	20	10
300	0.5	0.7	8	20	10
300	0.6	0.8	9	20	10
300	0.7	0.9	10	20	–
400	0.8	1.0	11	20	–
575	0.8	1.2	14	25	–
725	0.9	1.4	16	25	–
750	0.9	1.7	19	30	–
750	1.0	1.6	18	30	–
750	1.2	1.6	18	30	–
750	1.3	1.6	18	30	–
750	1.0	1.6	18	30	–
750	1.1	1.6	18	30	–
750	1.3	1.6	18	30	–
750	1.0	1.6	18	30	–
750	0.9	1.6	18	30	–
750	0.9	1.3	15	30	–
750	1.0	1.3	15	30	–
750	0.8	1.3	15	30	–
750	0.7	1.3	15	30	–
750	1.0	1.6	18	60	10
1,200	1.1	1.8	21	60	10

Appendix 2

Nutrient analysis per total recipe

The figures given below are the nutrient analysis for the total ingredients in each recipe and are the basic data for the 'per portion' figures at the head of the recipes. If the number of portions obtained from a recipe is altered the figures below should be divided by the adjusted yield to obtain the correct level of nutrients per portion.

Note: The % energy from fat will not differ with any change in recipe yield.

*The figure for sodium does not take into account any possible additions in the form of stock cubes, tinned or processed ingredients other than those incorporated into the food tables.

Salt stated in the recipes, or included in the sauces, is calculated in the sodium figures. If further salt is added to a recipe the sodium figure should be adjusted by 387 mg per 1 g of salt.

† represents a negligible amount of fibre.

Recipe index

The dishes in the book are listed alphabetically with appropriate page numbers.

Recipe	Page No.	Energy value kJ	Energy value kcal	% energy from fat	Protein (g)	Fat (g)	Carbohydrate (g)	Fibre †(g)	Sodium *(mg)
Alubias	136	13972	3316	29	240	109	365	169	10426
Apple dumplings	188	14913	3551	21	47	86	783	73	1448
Apple strudel	197	8792	2074	17	41	40	412	41	201
Arabian salad	175	1695	400	34	19	16	50	24	251
Armenian vegetable stew	119	6040	1435	31	102	50	160	48	2829
Avocado dressing	170	5563	1340	74	57	111	30	17	280
Baba ganouge	175	1147	266	35	9	11	36	20	42
Baked apple dumplings	188	14913	3551	21	47	86	783	73	1448
Baked filled potatoes	137	11831	2781	13	145	42	486	51	3284
Barbecue sauce	77	11719	2790	31	107	97	398	52	5671
Bean goulash	104	10225	2427	32	258	88	162	49	2757
Bean salad	176	3631	853	5	62	5	149	64	231
Beef carbonnade	132	8126	1937	35	217	77	63	23	3144
Beef galettes	155	6309	1494	25	168	42	117	38	3402
Beef olives	105	12026	2863	28	242	91	282	37	4847
Beef spread	173	5786	1378	35	254	46	8	1	915
Beef stew	102	9483	2252	31	253	80	137	53	3227
Beef and vegetable pie	111	10604	2505	24	200	70	286	37	3587
Blanquette of lamb	108	16630	3936	25	285	111	477	39	3588
Bolonaise sauce	79	26561	6324	35	698	250	341	56	8342
Bolonaise and vegetable sauce	80	22093	5260	34	501	205	376	74	8096
Braised millet—Persian style	131	12158	2907	28	91	91	480	36	2209
Braised rice—Indian style	129	12565	2962	12	85	41	601	65	2062
Brandy sauce	215	3582	840	4	364	13	627	1	5520
Bread pudding	214	6950	1645	30	75	57	222	19	2009
Broths—see mutton broth									
Brown beef stew	102	9483	2252	31	253	80	137	53	3227

Recipe	Page No.	Energy value kJ	Energy value kcal	% energy from fat	Protein (g)	Fat (g)	Carbohydrate (g)	Fibre †(g)	Sodium *(mg)
Brown mutton stew	102	11103	2642	41	258	121	137	53	3489
Brown onion sauce	78	15207	3621	34	118	139	506	92	5135
Brown rabbit stew	102	9513	2262	29	269	74	137	53	3279
Brown sauce	76	8971	2127	32	78	79	296	24	5474
Brown veal stew	102	8903	2112	25	261	61	137	53	3716
Carbonnade of beef	132	8126	1937	35	217	77	63	23	3144
Cauliflower soup	86	6506	1522	6	108	10	268	55	4891
Channa dahl	118	8472	2001	27	106	63	269	77	2199
Charlotte montreuil	221	7957	1871	14	86	31	332	5	1396
Charlotte russe	220	7885	1856	15	88	31	326	3	1404
Cheesecake	223	10470	2485	33	124	94	304	3	3335
Cheese sauce	74	19619	4643	27	351	141	524	24	11102
Cheese spread	172	10191	2426	48	252	132	10	1	6426
Cheese and vegetable spread	172	8599	2047	44	196	103	17	5	5135
Chicken chasseur	143	11720	2789	34	364	108	52	13	1934
Chicken khoreshe	150	22841	5416	24	363	161	668	52	1245
Chicken spread	173	8255	1965	32	254	71	8	1	1196
Chick-pea dressing	168	3967	933	16	63	17	142	41	423
Chilli con carne	106	9876	2340	28	227	75	202	73	2866
Chinese pork	107	14117	3356	28	236	105	379	15	4728
Chocolate gâteau	203	17124	4053	27	82	124	696	13	3246
Chocolate mousse	219	5204	1228	24	81	35	157	–	1288
Chocolate sauce	215	17511	4115	5	375	26	634	1	6090
Choux pastry	191	5617	1337	52	45	79	120	5	353
Clam chowder	94	7401	1765	36	82	73	201	47	7026
Coffee gâteau	204	18601	4286	33	94	167	639	34	2770
Cornflour sauce	215	3031	707	2	34	4	150	1	541

Cornish pasties	190	14839	3557	35	128	141	463	69	5834
Cottage and farm pie	112	8797	2074	20	161	47	268	46	2731
Coupe—Alexandra	226	469	112	24	3	3	19	1	43
Coupe—Edna-May	227	458	109	24	3	3	18	1	44
Coupe—Venus	227	417	99	27	3	3	16	1	42
Cream Chantilly	217	8432	2043	89	11	203	46	–	197
Currant sponge pudding	203	13622	3243	10	67	36	696	22	2771
Curry sauce (European)	81	12712	3027	35	73	119	448	24	5366
Curry sauce (Indian)	82	7239	1723	32	78	63	225	20	1935
Custard sauce	215	3031	707	2	34	4	150	1	541
Devil sauce	77	10214	2432	29	87	80	325	47	5634
Egyptian bread salad	178	2825	666	23	28	18	106	32	1158
Escalopes of veal	145	12411	2945	23	320	78	231	68	3738
Eve's pudding	200	15445	3677	9	68	36	809	28	2727
Farmhouse casserole	133	8904	2114	28	251	67	134	38	3676
Fenugreek soup	93	9437	2254	36	112	92	260	20	4847
Fish cakes	156	4402	1035	13	136	14	96	16	2713
Fish pie	125	13946	3302	25	274	94	362	48	2799
Fish Suchet	121	6926	1637	12	278	23	65	20	2122
Fish Véronique	122	14546	3439	25	551	100	71	9	4426
Fruit flan	184	10360	2456	32	46	90	389	16	471
Fruit gâteau	205	19344	4585	31	89	161	741	31	2777
Fruit kebabs	211	3707	873	11	10	11	195	35	36
Fruit mousse	220	4271	1008	19	69	22	153	–	503
Fruit pie	186	8123	1915	20	22	43	384	26	656
Fruit sauce	216	1440	340	0	8	–	81	71	28
Fruit tart	187	10534	2487	22	32	64	475	29	858
Fruit trifle	201	11674	2779	25	55	78	489	16	1807
Garlic dressing	165	10601	2571	94	18	122	20	2	3915
Gazpacho	92	10557	2536	73	40	210	131	45	4222
Guacamole	169	5324	1284	78	34	113	34	15	174

Recipe	Page No.	Energy value kJ	Energy value kcal	% energy from fat	Protein (g)	Fat (g)	Carbohydrate (g)	Fibre †(g)	Sodium *(mg)
Ham spread	173	6003	1429	32	162	52	8	1	11124
Herb sauce	164	9787	2367	92	26	243	21	14	1489
Hummus	168	4899	1175	35	67	47	129	37	125
Hungarian chicken	142	21242	5058	24	424	139	533	48	2005
Ice-cream—vanilla	225	12953	3076	35	108	122	410	1	1668
Ice-cream—raspberry	225	13373	3176	34	112	122	432	31	1680
Impératrice rice	222	6846	1610	13	71	25	295	4	809
Irish stew	109	11499	2723	30	245	90	250	45	3231
Kofta kebabs	157	17345	4109	28	232	131	536	34	2893
Lambs kidneys in Madeira sauce	147	7305	1739	35	182	68	86	15	3274
Lemon meringue flan	185	13682	3258	22	47	80	556	12	1220
Lentil dahl	118	7947	1878	17	122	37	281	60	2191
Lentil soup	83	11195	2637	3	197	8	472	123	690
Macaroni and lentil bake	101	9124	2172	23	102	57	342	52	594
Minted lamb kebabs	158	18460	4377	28	280	138	559	36	3148
Mixed pudding	213	5537	1297	3	52	4	283	8	685
Moroccan carrot salad	178	1007	236	0	7	0	55	29	951
Moussaka	110	9890	2355	27	567	73	226	88	3281
Mulligatawny soup	88	9839	2332	32	71	86	339	30	5176
Mutton broth	84	2853	669	4	21	3	149	34	4490
Nan bread—plain	206	9774	2303	16	69	43	438	19	3343
Nan bread—wholemeal	206	9106	2153	19	587	47	368	50	3364
North sea fish stew	124	17477	4143	22	310	103	500	53	3234
Orange dressing	170	2059	494	14	41	8	69	–	612

Item									
Pastry									
choux	191	5617	1337	52	45	79	120	5	353
pudding	192	9804	2334	40	40	104	327	14	2368
short savoury	181	11550	2756	49	39	154	322	14	1375
short sweet (1)	181	12222	2910	46	39	154	364	14	598
short sweet (2)	182	9678	2295	33	40	86	364	14	1396
wholemeal	183	9038	2156	43	53	105	265	38	2339
Pasta									
plain fresh	196	7691	1823	31	69	64	260	21	2140
wholemeal fresh	196	7343	1743	29	58	58	263	38	2068
Pasta ratatouille	116	5156	1219	29	100	40	123	50	2249
Pasta with bolonaise and vegetable sauce	95	16588	3950	20	275	89	557	56	3282
Pasta with pine nuts	96	14008	3334	23	111	85	567	63	4936
Pasta and potato bake	100	12861	3036	21	134	74	488	52	2896
Pasta salad	179	4224	1006	9	46	10	192	45	289
Pineapple dressing	166	10327	2500	88	22	245	55	5	1501
Piquant sauce	79	10214	2432	29	87	80	325	47	5634
Pizza									
plain dough	207	9067	2134	15	63	36	415	19	2113
tomato and vegetable	208	12363	2927	23	134	77	452	34	4045
wholemeal tomato and vegetable	208	11724	2784	25	152	81	383	65	4076
fruit	210	7561	1783	21	32	43	339	13	2488
Potatoes filled baked	137	11831	2781	13	145	42	1053	51	874
Potatoes roast	139	6319	1484	13	32	21	312	32	105
Prawn spread	174	5795	1380	10	209	25	8	1	14809
Ratatouille	115	2721	639	28	28	20	97	35	2046
Rice pudding	212	4962	1160	2	50	2	250	3	657
Rice salad	177	5410	1270	6	29	9	285	29	541
Riz pilaff	128	10414	2457	16	42	47	498	28	1980
Roast potatoes	139	6319	1484	13	32	21	312	32	105

Recipe	Page No.	Energy value kJ	Energy value kcal	% energy from fat	Protein (g)	Fat (g)	Carbohydrate (g)	Fibre †(g)	Sodium *(mg)
Salad dressing 1	162	21203	5143	94	36	541	40	4	7829
Salad dressing 2	163	23140	5600	88	66	549	109	14	7885
Salat	177	678	158	10	10	2	27	15	89
Scones	209	5326	1256	23	30	33	225	8	2483
Seafood mélange	123	13877	2800	12	207	46	414	33	5049
Seafood and potato pie	126	13705	3245	25	335	93	285	44	6248
Seafood risotto	130	13147	3108	13	147	46	560	33	7482
Sicilian ratatouille	116	5855	1388	38	73	60	151	46	2882
Sponge cake	199	12343	2903	11	65	36	618	13	2707
Steak and vegetable pudding	194	14819	3521	35	212	142	368	28	5117
Steamed apple pudding	193	13206	3127	29	45	104	534	39	2445
Stir fried chicken with mushrooms	148	17819	4254	20	281	98	560	14	4344
Stir fried chicken with peppers	149	18139	4292	20	281	101	599	41	4343
Tandoori chicken	159	7335	1749	29	271	57	39	2	3274
Thickened vegetable soups —see cauliflower soup									
Thousand Island dressing	165	9648	2339	93	17	243	22	3	1278
Tomato sauce	81	12853	3049	32	131	111	406	38	5572
Tomato soup	89	10015	2374	31	127	84	297	47	6792
Tomato soup (fresh)	90	9630	2276	26	85	67	353	98	5700
Tournedos with mead	144	11979	2825	29	383	92	102	29	2280
Tsatziki dressing	166	1029	260	14	22	4	37	2	347
Turkish shepherd's salad	179	1753	408	3	12	1	94	25	413

Vanilla mousse	218	4423	1041	19	70	22	150	–	718
Vegetable burgers	154	7292	1736	13	73	26	303	53	1509
Vegetable curry	117	7091	1675	22	64	43	276	79	913
Vegetable dumplings	97	13502	3415	33	86	120	475	61	1682
Vegetable hotpot	134	6966	1642	19	73	35	278	61	3527
Vegetable hotpot with cheese	135	8753	2084	23	113	54	303	66	4363
Vegetable lasagne	99	13654	3238	30	169	111	414	86	5333
Vegetable quiche	189	8105	1932	45	64	98	211	12	1477
Vegetable raita	167	2096	496	10	39	5	79	26	835
Vegetable spread	174	2598	617	26	35	18	84	24	1008
Venison in red wine	146	14157	3371	32	440	123	109	7	1886
Vichysoisse	91	8599	2027	7	65	16	435	84	4738
Victoria sandwich	200	13682	3257	10	66	36	701	14	2721
White sauce									
Béchamel	73	19099	4519	35	235	180	523	24	8179
velouté	75	8399	1993	35	66	79	273	24	4619
Wholemeal scones	210	5050	1196	27	39	35	194	22	2501
Yorkshire pudding	140	2623	618	18	30	13	102	8	1394

Index